ELECTRICAL PROJECT MANAGEMENT

SAM MELAND, P.Eng.

ELECTRICAL PROJECT MANAGEMENT

McGRAW-HILL BOOK COMPANY

*New York   St. Louis   San Francisco   Auckland
Bogotá   Hamburg   Johannesburg   London
Madrid   Mexico   Montreal   New Delhi
Panama   Paris   São Paulo   Singapore
Sydney   Tokyo   Toronto*

Library of Congress Cataloging in Publication Data

Meland, Sam.
    Electrical project management.

    Includes index.
    1. Electric engineering—Management.   2. Industrial
project management.   I. Title.
TK441.M4   1984   621.3′068   83–957
ISBN 0–07–041338–X

2345678910   VBVB   898765

ISBN 0-07-041338-X

The editors for this book were Diane Heiberg and Janet B.
Davis, the designer was Elliot Epstein, and the production
supervisor was Teresa F. Leaden. It was set in Caledonia by
Com Com.

Contents

Introduction   vii

1   Functions of a Project Manager   1
2   Activities Common to All Operations   11
3   What the Field Should Know about Estimating   19
4   Costs Are Higher Than You Think   29
5   Change Orders   41
6   Your Contract and Impact Costs   49
7   Claims   61
8   Work Force and Worker-Hours   77
9   Productivity   91
10 Prefabrication   105
11 Purchasing   127
12 Material Storage and Handling   151
13 Tools   165
14 A New Look at Temporary Wiring   177
15 Job Office Procedures   187
16 Procedures for Managing a Cost-Plus Project   213
17 Managing and Controlling with Computers   231
18 Progress Billing and Cash Flow   253
19 Profiles and Pitfalls of Typical Projects   263
20 A Positive Professional Attitude   275

Appendix: Conversion Factors   281
Index   283

Introduction

Let's face it. People in the construction industry are technically oriented. You tend to regard modern management techniques and paperwork as secondary activities. You pay lip service to their importance in your operation, but do you really practice what you preach?

You are very interested in every new development in construction technology and tools—but how much effort do you really make to keep abreast of the new developments in management techniques and to apply them in your day-to-day work? Yours is a dynamic, fast-paced, high-cost industry. Keeping abreast of the competition and meeting tight construction budgets, while still making a profit, require a planned, coordinated, efficient style of work. Yet there is a surprising amount of sloppiness in the management of construction operations. The jobs run you as often as you run the jobs.

To make a profit despite the ever-increasing cost and complexity of doing the work requires that you control your cost and schedule for on-time performance. It requires a style of work that compensates for human nature and channels it into productive paths. It requires uncomplicated procedures that are easy to enforce and that flow from your normal activities.

Most of us drive to work. If we can drive one car, we can drive them all. Automobile manufacturers have made driving easy by standardizing components and procedures. They have simplified the task of driving by reducing it to a reasonable number of basic operations. They have made these operations repetitive and habitual. They have built in reliability and predictability. There is a lesson to be learned here.

Construction contracts can be thought of as vehicles for transporting a fair profit to the company at the end of the project journey. The procedures for management and control must be consistent, standardized, and common to all jobs. A project manager should be able to get into the driver's seat and run any job, because the management operations and procedures are familiar. They have been repeated so often that they have become second nature. Signals flow smoothly back and forth between the manager and head office because all are following the same game plan and the plays are thoroughly understood. Controlling the jobs means setting and meeting targets, both for production and for costs. Meeting targets, in turn, requires effective scheduling, job-coordinated material releases, and the channeling of updated information to the working crews.

Systems are important, but without the right people they won't work. Finding and keeping good workers involves personnel evaluation, training, and motivation. Planning is also required to compensate for lapses and to cut down lost time. As in any sport or competitive game, you cannot focus on your own moves solely. You have to be aware of the moves of the other players, and you must look and think ahead. Construction projects are dynamic, and job events affect costs. You have to be conscious of these impact factors so that you can protect yourself. You should not be fatalistic about adverse job factors; you should protect yourself against their effects. However, in order to protect yourself adequately, you must understand the nature of the problems that you are up against.

To become aware of problems before they affect the work, head office staff and project managers must set up regular inspection and reporting procedures and staff meetings. Actual productivity must be compared on a regular basis to acceptable norms, and corrective action taken, if necessary, before too many worker-hours are lost. Your biggest challenge is to manage workers and worker-hours in a way that improves productivity and meets your targets. Everybody talks about productivity, but you and your industry must take the specific actions needed to improve it. The solutions are complex—yet perhaps simpler than you think.

Every project undergoes revisions. Standard procedures are required to keep track of these changes and to produce signed documents for billing. It helps to have a good understanding of the nature of costs so that they can be properly explained to your customers.

Much loss of time, effort, and money results when you don't do the commonsense things that flow from good management. You have to find the ways that will enable you to generalize and learn from your experience, and the time for doing so. By sifting through your experiences, you will come up with procedures that are simple, straightforward, and consistent and that suit your operation.

This book is intended to provide basic reference material for project managers and job foremen and the coordinators and office personnel with whom they deal. There are no special schools for producing and educating project managers and foremen. They usually come up from the ranks, and are singled out for added responsibility because of special aptitudes and leadership abilities. Most contractors have this background, as well. These contractors may be large, small, or one-person operators. All of them, and all personnel who run construction jobs or supervise other workers, require management tools and techniques as surely as they require production tools and techniques. It is hoped that by directing attention to the routine but important matters of project management, this book will be a valuable tool which will help them all to better discharge their managerial responsibilities and to better run their businesses.

Although this book is directed toward electrical contractors, it should be useful to other trade contractors and to general contractors who have similar problems. It should also be of interest to architects and engineers who interact with contractors, and who have a professional duty to truly understand the practical problems contractors encounter in the course of building a project.

Sam Meland

ELECTRICAL
PROJECT
MANAGEMENT

1
Functions of a Project Manager

Becoming a project manager or a foreman is very often like becoming a parent. It is assumed that all the required characteristics and functions will develop or be acquired naturally. This approach may have been adequate in the past, but not in today's market. The jobs are too complex and the cost of doing the work is too high and too risky to leave the outcome to chance.

You have to learn the lessons that other industries have learned or are learning: that in order to survive you must develop professional managers who truly understand the nature of your industry. You have to develop the educational institutions to produce these professional personnel. As an analogy, think back to the past, when the power delivered to projects was small. Under those circumstances, no one paid much attention to the rupturing capacity of the electrical equipment that was installed. Today, when the power networks are immense and the amount of rupturing current that can be fed into even a small fault is extremely large, everybody is concerned about rupturing capacity. When even a small fault will result in rupturing currents large enough to cause severe damage and endanger life, it had to become a factor of great concern.

The same is true of construction projects. When the projects were small and wage rates very low, the construction industry was not as aware or concerned as it is today about low productivity, job impact costs, and the need to plan and schedule the work. Nowadays, when the failure of even a modest contract has the rupturing capacity to knock a contractor out of business, both the contractor and the industry must become keenly aware and knowledgeable of what must be done to prevent that from happening. *Project managers are key persons in this regard*. Their functions are as follows.

PLAN THE WORK

Every job eventually gets done. The challenge, however, is to get it done within budget and on target. This will not happen if jobs are allowed to develop on their own, on a day-to-day basis.

A project gets done as a process of starting and completing segments of the work—commonly called *operations.* Some operations are carried out simultaneously; others follow one another in sequence. There is interdependence, sometimes of a critical nature, between operations, and also with the operations of other trades. Updated design information and the correct material must be available in the right place at the right time, so as not to disrupt the operations. Only through planning, coordination, and scheduling can a job be brought off in an efficient and economical manner.

Start by breaking down the job into operations. Allocate the operations into the time frame when they are likely to be done, after checking with the general project schedule and with the other trades. Don't be afraid to make assumptions if hard information is not available. It is better to start with a rough picture of how the job is likely to develop than with no picture at all. The resulting bar chart or operations network will, no doubt, be updated many times as hard information becomes available.

Determine the point in time when an operation must be completed. By looking back from this point, establish the dates when key activities must take place: purchase order placed, shop drawings submitted, coordination worked out, material released and delivered on site, operation started, and operation completed. Repeat the process for all operations. All this is done in advance, when there is still sufficient lead time to account for all the activities that are required. The key to your planning is to ensure that the required material arrives at the job on time, so as not to disrupt or delay the work. Countless worker-hours* are wasted while waiting for material, or while working around an operation until the correct material arrives.

Planning is a chain that is only as strong as its weakest link. You have to see to it that every link is strong and will carry its weight. Regular discussions with your involved personnel and with your suppliers and manufacturers are essential. Review the program, and get feedback from those overseeing the other links in the chain. Establish critical dates when material must be delivered and operations must be completed.

An important purpose of planning is to ensure that you don't repeat past mistakes. Human nature being what it is, people tend to make the

---

*1 worker-hour = 1 man-hour. As a unit of labor, the terms are interchangeable.

same mistakes rather than to learn from having made them in the past. You must make the effort consciously to learn from feedback on past jobs, so that you can include this experience in your planning. You won't get real feedback on job performance unless you ask the right questions —either in face-to-face conversations, in meetings with well-thought-out agendas, or by seeking regular reports.

In your planning, be aware of the following pitfalls. Too much time is often wasted at the beginning of the project, before the purchase orders for the main equipment and materials are placed. This can cause problems down the line, when critical material does not arrive at the job on time; disruptions to the schedule follow, and costly catch-up work must be done out of sequence. A contract is like an airplane flight, in which the critical points are takeoff and landing: in a contract, the critical points are starting the job and finishing it. Very often the job is overstaffed at the beginning, and more worker-hours are lost at this point than can ever be made up. At the finishing stage worker-hours are also lost, because bad planning and bad job management result in over-staffing to pick up lost time and to correct defective work.

Time is not given sufficient status as a factor in construction costing. All parties concerned are very slack in producing schedules. Without schedules there can be no effective planning. You have to discipline yourself to schedule your work, to insist that your suppliers schedule their deliveries to suit, and to insist that you get a proper schedule from the general contractor or the contract manager. A major reason for cost overruns is poor scheduling, which, in fact, means poor planning.

CONTROL THE COSTS

In the field, the job will get done—but often at a cost higher than budget or higher than necessary, because the people involved are not cost-conscious.

You all know that people involved in sports are performance-conscious. Every runner of the mile tries to approach or beat the four-minute mile. You are bombarded with batting averages, earned-run averages, and dozens of other statistics, all of which give you a measure of performance in various sports. Unfortunately, this attitude is not representative of the general approach of construction personnel to their work. There is not a gut understanding or acceptance that a given operation must be done in the minimum number of worker-hours necessary, based on the particular job conditions and an acceptable tempo of work. Labor worker-hours are very well understood by the estimator who prepares the bid, but not by the foreman who supervises the work. There are scores of handbooks of labor units with factors for height and job conditions. However, this information has not sufficiently filtered down to the installers in the field or their supervisors, nor is it a factor during labor negotiations. Your field personnel are just not sufficiently aware of how you arrived at your labor estimate. Part of controlling the cost is to demand of your field people that they perform to acceptable standards. You have to educate your field personnel so that they acquire a realistic understanding of cost.

The work must be planned and coordinated to achieve this end. Refer to the worker-hour data contained in the estimate and to previous job experience to ensure that the project is not overstaffed. If a project is progressing slowly, an oversized crew will slow down to fit the resulting work load. Everyone will appear to be busy, but the cost will be higher than budgeted.

The foremen should have, or be given, a clear idea of a realistic worker-hour target for the operations under their control. They should set realistic worker-hour targets for their crews and check the performance accordingly. Talk worker-hours to your installers until labor units and performance standards become as familiar to construction personnel as performance standards are to the people involved in sport. The attitude of the installers is an important factor in performance, but so is the attitude of management. Worker-hours that should be spent on productive work must not be squandered on nonproductive activities, such as looking or waiting for information, material, or tools.

Your foremen should understand and feel that they are a strong link in the chain of management to control the costs. This means that they are part of the coordination process. Since they are most aware of job conditions, they should help to come up with the most efficient and economical methods of running and fastening the raceways and the other materials for which they are responsible. Walk through any job and you will see dozens of examples of uneconomical or wasteful fastening and installation arrangements. You should review the work continuously and insist that your installers come up with economical, smart solutions.

Plan to prefabricate as many assemblies as possible in your shop or under shop conditions. This is the style of work that will dramatically improve productivity. Keep in mind that you lose a good part of your worker-hours in the installation of conduit. Single out this operation for special control. As goes the conduit installation, so goes the job. Place some of your best workers in this operation, so that they will set the pace for the others to follow.

Too many hours are lost in extended break periods, lunch periods, punching in, and punching out. Some of the lost time is due to job conditions, but much of it is due to poor attitude or poor planning. In both cases, the project manager must act to rectify the situation. Arrange to locate the shacks as close to the work areas as possible to cut down time lost going there and back for break periods. These shacks should be portable so they can follow the work. In the case of highrise buildings, where there is congestion at elevators and time lost waiting for elevators during morning start and break periods, arrange with the general construction management to stagger the starting time and break periods of the various trades, to minimize this loss of time.

MANAGE THE JOB

To be a good manager is to make things happen in accordance with a plan or goal. It requires the ability to analyze and set realistic targets, and then to follow through with periodic review to ensure that the targets will be met. It requires the ability to work with people, to understand them, to motivate them, and to lead them. Communication is the key. Before commencing an operation, look into the following activities with the foreman and the other people who will be involved.

Examine the details of the original estimate pertaining to the operation. Check to see if there are change orders that affect it. Review the coordination information regarding the best manner of installation, fastening, and noninterference with other trades. Review the work force required and the best available workers to do the installation. Establish realistic worker-hour targets, and measure performance periodically. Most of all, make sure that the latest information pertaining to an operation is available and is explained to the installers actually doing the work.

The work must be organized and administered along direct, simple lines. Delegate responsibility, and do not go over the head of the person so delegated. Authority must go along with delegated responsibility. Do not load your people with more responsibility or details than they can handle. Assign them to what they do best, and explain to them clearly what is expected of them. Sort out and correct duplicate functions, and take the time to review the performance of each person.

There should be a strong link between the field and head office. The field must receive the strong support that can only come from head office. Head-office personnel are immersed in the job costs and view them with sufficient perspective to exercise a controlling influence. The estimator should maintain a continuous liaison with workers in the field to apprise them of the changes to the contract. Because estimators are trained to think differently than installers, they are very often in a position to come up with labor-saving or material-saving ideas. The purchasing agents are in constant touch with suppliers, who introduce them to new products and ideas that may prove to be very useful in the field.

A well-managed, successful job is very often the result of a close, effective relationship and of interplay between field and head office. Make it a requirement of project management to hold regular staff meetings on the site, attended by head-office personnel such as the general superintendent, the contract manager, the estimator, and the purchaser. The sharing of information and the common focusing on job problems are vital ingredients of good management.

Another important factor is that all job communication with head office should be through one designated person at head office. This person is the contract manager, whose responsibility is to make sure that every job request is channeled to the right party, and to ensure that it is acted upon. You must work very closely with the head-office contract manager who is assigned to your project. Instead of having to contact many people at head office, none of whom has the specific responsibility to follow through, you communicate with your contract manager, who makes sure that the problems are looked after.

LEAD AND MOTIVATE

Your workers must be encouraged to do their best. This is no easy task since antiwork attitudes have been creeping steadily into your industry. The way to encourage workers is to make them feel involved. Take the time to review what has to be done, and give them a good idea of where they are heading. Ask for their opinions and ideas. The best motivation is self-motivation, behavior stemming from the drive of an individual to satisfy inner needs and goals. To instill such motivation, an individual must be made to feel that he or she counts. Encouraging such feelings will help to create a job climate in which your personnel will want to perform at their highest level. Set challenging, but realistic, productivity standards. Encourage good planning to achieve the targets which are set. Make sure that there is feedback about how the job is doing and what are the achievements and liabilities. Give your staff increased responsibility, together with the opportunity for personal control of situations under their jurisdiction.

Involving your people in the planning and achievement of job targets will stimulate their desire to work harder and be more persistent in their activities. Make them feel an important part of the team, and encourage their ideas. The desire for self-fulfillment is a great motivator. Once so motivated, your workers will respond to the challenge and to the pleasure they will get out of work itself, the feeling of responsibility, the sense of achievement, the recognition of a job well done, and the desire for self-advancement.

The old-fashioned project managers in construction were autocrats who didn't want their subordinates or workers to seek explanations. They demanded that the work should be done their way, with no questions asked. This type of leadership will not work in these modern, complex times. Construction personnel are better educated now and are influenced by egalitarian attitudes toward teamwork and job satisfaction. Management was slower than the unions to understand the power of teamwork and the satisfaction that comes from cooperation to achieve a common goal. These ends are not easily achieved on the jobsite. Workers are obtained from a common pool, and they identify more easily with their union or hiring hall than with your company. They may look on you as an adversary, rather than as a partner, in an industry in which you both want to keep competitive and healthy. The livelihood of you both depends on it.

As the manager, you must lead your workers to a more cooperative approach and a common pride in a job well done. Just think about it. How often do your workers get credit for a job well done? Do you really have teamwork on your job, or is your job ridden by personality clashes, territorial disputes, and conscious or unconscious sabotage of your work program? If your style is to plan the work without consulting or involving your personnel, don't be surprised if they don't put out for you or come up with good, labor-saving ideas. What you will get from your people under those conditions is only the minimum that they can get away with.

Good leadership and good motivation do not come from running a one-person show. They come from the realization that, along with good technical coordination, you must have good worker cooperation. Involve your people. Build up their confidence and self-esteem. Involve them in regular job meetings where they can review performance and learn from their mistakes. Encourage innovation, and create an environment where your people can prove themselves. There is tremendous potential in your personnel if you can only harness it. There must be rules, and you have to exercise discipline, but being a good leader you will use your disciplinary power fairly and with judgment. You can build up a good team approach and still enforce the rules. You cannot effectively drive your people; you must lead them.

ESTABLISH PERSONNEL REQUIREMENT

In order to establish the immediate and long-range personnel requirements of the project, review the estimate thoroughly, and determine the quantity of worker-hours for each operation. Review the general construction schedule, and draft the operations onto a bar chart showing the proposed start and finish of each operation. By incorporating the worker-hour information, work out the proposed personnel distribution graph. Distribute the noncritical operations to minimize undesirable fluctuations in the work force and to cut down the peaks and valleys which result in lost time.

If a crew has eight hours in which to do a four-hour activity, they will stretch the work to fill the available time. This can happen if the foreman hasn't planned what the crew should do next. Time is also lost because of slow periods between high-intensity operations, such as roughing in slabs.

The size of the gangs on the jobsites must be determined from a detailed analysis of the actual job schedule and must be based on the estimate of worker-hours necessary to do the given amount of work in the given time. Under no circumstances should a job be packed or overstaffed in order to compensate for bad planning, bad scheduling, or low productivity. Realistic and competitive production norms must be set, particularly for repetitive and critical operations, such as conduit installation, wire pulling, and fixture installation. You pay as much for incompetent workers as you do for those who know their business and want to produce. Set high standards. Insist on a good level of productivity and competency. Arrange the work by crews rather than by teams, so that a two-worker team does not end up doing the work of one worker.

At one time or another you will be subjected to great pressure to increase the size of your crew. There is often a conflict between the demands of the job schedule and your need to control the size of the crew in accordance with your targets. The size of the crew should always be based on careful planning and work programming. If there is an overriding principle, it is to keep the size of the crew as small as possible. The jobsite is like a sieve when it comes to worker-hours. Without planning and control, the worker-hours will run out as fast as you pour them in.

ESTABLISH QUANTITY AND TYPE OF SUPERVISION

Decide upon the number of foremen the job requires. Determine the maximum area that a foreman can cover and control effectively, and decide whether the responsibility should be by area or by system. Overlapping of foremen's responsibilities must be avoided, since it leads to confusion and conflicting orders. Foremen should spend most of their time in areas under their jurisdiction. They should have a set of pertinent plans on a worktable in their area. Consultations with other foremen or the project manager should be so arranged that workers are left alone a minimum amount of time.

Plan operations to obtain maximum efficiency from crews. When an operation is started, keep at it until it is completed. Starting an operation, leaving it for another, and returning to it at a later date make for a costly and inefficient style of work. All necessary material for an operation must be at the site before work begins. Any operation for which information is uncertain or material is lacking should be postponed until the missing elements are available. The foremen must have a thorough knowledge of the operation, the target worker-hours, and a coordinated work plan before committing their crews.

Loose ends should be picked up before the hectic finishing stages. Identifying and tagging circuits and installing box covers and plates at the same time as the wiring is being installed will save many worker-hours. Likewise, circuits should be identified and marked at the same time as the panels are dressed.

The field is not suited for the economical manufacture of equipment assemblies. The field operation basically calls for planned, coordinated installation of complete assemblies. A well-supervised job will utilize prefabrication wherever possible. There is a substantial saving in worker-hours when you order items to be made up in a well-equipped central fabrication shop, staffed by specialists using the best possible tools. An important by-product is that the foremen are forced to look at an activity before the work commences, to be able to order prefabricated items on time. This gives them the lead time to notice other items that may have been forgotten and that must be looked after, in order that delays may be avoided once the work has started.

Regular meetings should be held for the project manager and foremen to plan the work, check on progress, and review the worker-hours and the material requirements. These meetings should be well prepared and factual. They will help to determine the optimum crew size to suit the tempo of the work. Minutes should be kept and distributed to all concerned.

LEARN FROM EXPERIENCE

Learning is a painful discipline. It takes thought, application, time, energy, and some sacrifice of the many things that you would rather do. Every job is a gold mine of experience. You must understand the importance of digging for and refining the generalized experience and knowledge which you have gained. You will need them in order to identify the job factors, job-expense items, labor costs, and management techniques on which the continued success of your company so vitally depends.

You start to backslide when you become apathetic, when you procrastinate and avoid doing what has to be done. You can be sure that this is starting to happen when you spend too much time on unimportant details and lose sight of the overall picture.

If you don't look or search, you will not see or find anything new. If you don't keep records and write things down, you will forget them and not benefit from your experience. These are new times and you must learn the new techniques of management, which are:

1. Maximum use of prefabricated assemblies
2. A team approach at all levels to plan and coordinate the work, to reduce the impact of adverse job factors
3. Giving more thought and effort to reducing the time consumed by material handling and other nonproductive activities
4. More involvement of job personnel to increase productivity
5. More and better communication at all levels
6. Simple, standardized, and effective procedures for managing and controlling the jobs

2
Activities Common to All Operations

The management of labor is your most important function. How are the worker-hours spent during the average workday? A thorough understanding of the various activities that must be undertaken to complete an installation operation is essential, in order to minimize lost time and improve productivity. Such an understanding will emphasize the role and responsibility of other parties as well as the contractor in the struggle to increase productivity. The architect, the engineer, the general contractor, and the other trade contractors all affect the way in which the worker-hours are spent during the average workday.

All installation operations involve the following basic field activities to some degree.

ACTIVITY 1: STUDY AND CHECK PLANS

You begin an operation by first studying and checking the plans and specifications relating to it. This may also require that you study the mechanical, architectural, and structural drawings and finishing schedules, all of which may affect the work.

Your style of work should be such that the working crews are furnished with precise details and coordinated installation sketches or drawings before they start the operation. This will save study and checking time. Arrange to coordinate the runs of heavy conduit, tray, bus duct, and other such items on field sketches or drawings, and have them initialed by the general contractor and representatives of the affected subtrades in order to obviate possible interference. Review the fastening requirements, and work out the most suitable and efficient solution, particularly on repetitive work. On repetitive operations, every good idea gets repeated hundreds of times—and so does every poor one.

Lack of required information or incomplete plans will result in a loss of worker-hours. It is therefore essential that all studying and checking of plans be done and that missing information be obtained before a crew begins an operation. Take the time to summarize and organize the information from the original estimate takeoff sheets, the notes on plans, and the pertinent data in the specifications, and put these in a binder for easy reference.

Ascertain the extent to which brackets, conduit bends, equipment assemblies, fixture leads, and any other items can be prefabricated in your shop. This is an important part of the study and checking activity. Make up lists itemizing the types and quantities of fixtures, mounting frames, receptacles, and other materials that are to be installed in the various areas in order to expedite the work and cut down on lost time, particularly in the finishing stages of the work.

ACTIVITY 2: ORDER MATERIALS AND TOOLS

Material that is ordered late and, consequently, arrives late on the jobsite is a major contributor to the loss of worker-hours. There is a further loss when material is ordered in a sloppy manner, with the result that incorrect material is eventually delivered on the jobsite. It is very important to order material on time and to be precise in ordering it. You must not automatically rely on the head office or the supplier to know as much about job requirements as you do. Specify the dates when you require the material to be delivered to your job.

Establish the best method of shipping—for example on pallets or bundled—and what tagging and identification will be required. Whenever possible, order wire to be multiple-reeled and color-coded. Group your runs on the largest size of reels that you can easily handle. Incorporate job requirements into the orders. Items such as panels should be identified by specifying whether they should be flush or surface-mounted, whether the service entrance should be at the bottom or top, what size the lugs will be, what the required number of knockouts is, and what are other features that should be incorporated in the manufacturing stage to speed up installation in the field.

The major equipment, auxiliary equipment, lighting fixtures, conduit, and wiring are ordered by the purchasing agent at the head office. However, the jobsite foremen are closely involved in the ordering activity. They must release the material that they require for the operations under their control in sufficient time to ensure that it is delivered to the working crews when they need it for installation. They must also order all the miscellaneous material, such as fastening devices, brackets, nuts, bolts, and pulling compound. Failure to do so in a timely manner will delay and hinder the work.

Effective ordering can only flow from effective coordination and planning of the work. Last-minute, panicky ordering of materials results when the job is running you and not vice versa. Your foremen must work closely with the job storekeeper, giving the storekeeper sufficient lead time to prepare and deliver the material on time. Arrange to bulk-order repetitive items in manageable quantities, so as to avoid a multitude of small orders for the material.

An important part of the field ordering of material is the ordering of prefabricated assemblies. These must be ordered with sufficient lead time for prefabrication in the shop. Since there are many demands for service and delivery from the prefabrication shop by the various jobs, you will have to allow sufficient time to have your order fulfilled in accordance with your requirements.

ACTIVITY 3: RECEIVE AND STORE MATERIAL AND TOOLS

Many worker-hours are consumed in receiving and storing material and tools, both in the job warehouse and in the various areas where the work is being carried out.

Every item of material which arrives on the jobsite must be carefully checked by the storekeepers. They acknowledge the receipt of the material by making out a receiving report, which they send to the office along with the respective shipping slips. The materials must be stored in a neat and efficient manner, so that they are readily available for the working crews. Arrange for bulk material, such as conduit, wire on reels, and fixtures, to be stored on the floors or in the areas where the work is taking place. Prefabricated materials should be stored in mobile baskets, on conduit buggies, and on pallets that can be moved with a pallet lifter, as close to the work as possible.

Provide the crews with a sufficient quantity of lockable storage boxes for storing small materials and tools. Spend enough time planning the storage of materials and tools on the jobsite, so that you will reduce the time lost in moving materials to the point of usage or in workers' wandering around looking for materials.

ACTIVITY 4: MOVE TO POINT OF USAGE

For the workers to do the installation, they have to receive the required material. The activity of moving material and tools to the point of usage must be carefully planned and organized. Just think of the worker-hours that are lost on your everyday job when the installers mark time while waiting for materials. Productive work is disrupted in order to search out or bring over the required materials.

Don't take this activity for granted. Think it out carefully, and implement an effective delivery system. This will have a positive impact on the work and result in improved productivity. Problems relating to the vertical and horizontal movement of materials must be carefully worked out by the foremen and the storekeeper, in order to move the material to the point of usage on time, without expending excessive worker-hours.

If enough locked storage boxes are maintained throughout the job, the time required to move material and tools for daily use will be greatly minimized. The use of adequate storage facilities will reduce the time lost in moving materials to the point of usage and will cut down on workers' wandering around looking for materials or having to visit the stores.

ACTIVITY 5: TOOL UP

To obtain the maximum output from the productive worker-hours, you should supply your installers with the best possible tools, scaffolding, and material-handling equipment. At the same time, your foremen and installers should be taught to respect tools, to use them safely and properly, and to clean and maintain them.

Construction is becoming more and more mechanized. Assembly-line techniques, work rationalization, and other techniques to increase productivity are based on sophisticated and specialized types of tools. However, there is a time and place for every type of tool. The cardinal principle should be to use the simplest tool that will do the job. Discretion must be used in the choice of tools. Often, expensive pieces of equipment, such as mechanically operated scissor lifts, are used for routine tasks because they happen to be available on site. This equipment is subject to stringent safety regulations governing its operation, which makes it expensive to use in situations where regular scaffolding would be a much more practical and economical choice. Overtooling, like overstaffing, can be wasteful.

In tooling up, you must not compromise on the quality of tools that you supply to your installers. Choose tools that are rugged and simple to operate and that will stand up to the rigors of the jobsite. Tools that frequently break down, are complicated to operate, and are subject to stringent regulation have their time and place, but they should be used only when they will be most effective in getting the job done. The installers will stand around while a tool is repaired or replaced, and worker-hours will be unnecessarily wasted. Don't be penny-wise when you are tooling up.

ACTIVITY 6: MEASURE AND LAY OUT

Before the actual installation activity can proceed, the work must be physically located and marked out. To cut down the worker-hours consumed by this activity, it is important that the major investigative work should already have been done—that the plans should have been checked and the coordination worked out. All measurements should have been checked and verified, and interference problems worked out.

Obviously, many worker-hours will be saved in the field if dimensions, locations of equipment, mounting heights, and future obstructions are shown on field sketches or coordinated drawings. To get the work done properly and on target, the foremen must study the information which has been coordinated and summarized from the pertinent drawings and specifications and must impart this information to the working crews.

Every one of the nonproductive activities described to this point is essential and must be accomplished before you can commence the productive installation activity. They all consume worker-hours. If the work is planned and coordinated, the number of hours consumed in these nonproductive activities will be greatly reduced. This reduction will be analyzed and quantified further on.

ACTIVITY 7: INSTALL THE WORK

A common fallacy is to think that the labor cost of an operation consists of the worker-hours consumed in the actual installation activity itself. This activity, no doubt, consumes a large portion of the labor, but the total cost consists of all the worker-hours consumed by all nine of the activities listed in this chapter, which together constitute the complete operation. Every effort must be made to decrease the time spent on nonproductive activities, so as to expand the amount of time available for productive installation activity. Staff your jobs with the most qualified installers possible, and institute programs to train them and upgrade their skills. The high cost of labor is forcing the trade to adopt production-line methods and to utilize prefabrication wherever possible. Sustaining a favorable wage level requires a planned positive work climate, a desire to think and use good judgment, and a will to work.

Productive worker-hours are lost because of poor planning or no planning. Installers are often not properly informed about the details of their work. There is often a sloppy attitude on the part of all concerned toward tying up loose ends and thus preventing future deficiencies.

When you examine the productive activity, there are two factors which you must act on and control:

1. You must expand to the maximum the amount of time that can be made available for this activity.
2. You must increase to the maximum the efficiency of the work that is done during this activity.

ACTIVITY 8: CLEAN UP

By day's end, or whenever an operation is completed, excess materials and tools must be returned to the locked storage boxes or to their respective storage locations. All refuse left by the work must be cleared up. This consumes worker-hours. It is essential that your field people should understand the importance of cleanliness and orderliness on the job. Good housekeeping will save worker-hours.

ACTIVITY 9: BREAK PERIODS

During the day, when activities 1 to 8 are taking place, worker-hours are consumed by the following:

• Break periods
• Late starts and early finishes
• Early lunches and late returns
• Personal comfort breaks and other rest periods

Even if the field people lived up to the letter of the labor contracts, they would still consume 20 percent of the working day on activity 9. As it is, the break periods are stretched, so that they often consume 30 percent of the worker-hours. With proper planning this activity can be reduced to a minimum. If possible, stagger break periods to decrease the load on hoists or elevators. Place toilets and workers' shacks as close to the work areas as possible, and enforce discipline regarding the length of break periods.

Activities 1 to 9 constitute the total cost of the labor operation, but if deficiencies or breakdowns should develop during the course of a job, additional labor will be needed to correct the deficiencies and to make good the work covered by the guarantee period. Deficiencies and breakdowns are usually counted as job expenses, and as such are discussed in Chap. 4.

By analyzing the activities that make up a labor operation, you will see that the working day breaks down into four major activity categories:

• Indirect labor
• Material handling
• Productive labor
• Break periods

Each of these major categories consumes worker-hours. It is imperative that you know what percentage of the working day is consumed by each, so that you can institute the programs necessary to increase productivity.

Productivity studies carried out in the United States and Canada have established the percentages of total worker-hours that are consumed by the various activities that make up labor operations. Obviously, job conditions vary, and the results obtained from these studies range from minimums to maximums. These values and their averages are shown in Table 2.1.

Even if you managed to get your installers to work harder and faster, the overall productivity would not increase greatly because the increased tempo would apply only to the 35 percent of the worker-hours that are, on the average, available for direct work activity. It is evident, therefore, that the challenge must be to find ways to reduce substantially the worker-hours consumed by nonproductive activities, in order to expand the time available for productive work.

You should plan and program your work to reduce nonproductive labor activities to 10 percent, material-handling activity to 10 percent, and break periods to 20 percent of the total activity. This will allow you to maximize the time for productive activity to 55 percent of the total activity, which should be a target for the industry. This is a challenge that must be undertaken not only by the contractors, but also by the unions and by the consultants whose design drawings, specifications, and attitudes have an important impact on site productivity. This topic will be dealt with in later chapters.

**TABLE 2.1   LEVELS OF PRODUCTIVITY, BASED
ON PERCENTAGES OF TOTAL WORKER-HOURS
CONSUMED BY VARIOUS WORK ACTIVITIES**

ACTIVITY CATEGORY	ACTIVITY NUMBER	PERCENT OF TOTAL WORKER-HOURS CONSUMED BY LISTED ACTIVITIES FOR EACH PRODUCTIVITY LEVEL		
		LOW	AVERAGE	HIGH
Direct work	7	15	35	55
Indirect work				
Study and check plans	1			
Order material and tools	2	20	15	10
Tool up	5			
Measure and lay out	6			
Material handling				
Receive and store	3	30	20	10
Move to point of usage	4			
Break periods				
Late start and early finish				
Early lunch and late return	9	30	25	20
Personal comfort breaks				
Coffee breaks				
Miscellaneous				
Clean up	8			
Power failures		5	5	5
Accidents	Other			
Other unknown				

3
What the Field Should Know about Estimating

Construction is one of the few industries in which you must price your product before you actually produce it. Since estimating contains an element of forecasting, or attempting to predict events that have not yet happened, it is not surprising that it is often surrounded by misconception, myth, and even mysticism.

LABOR UNITS

As soon as this or that favored or published estimating system or system of labor units is applied to a takeoff, many otherwise logical and practical construction people, as well as many associated professionals, automatically assume that the resulting cost represents God's truth.

How can any set of labor units—which at best represent averages of averages of data collected from many different jobs, in many different places, under many different conditions—possibly fit exactly the particular conditions and requirements of any given work or project? How can any set of labor units be applied, unless the user knows how these units were constructed, what they actually include, and under what job conditions they were taken? Every estimating system is only a guide which experience, judgment, and the governing job conditions must inform.

The only valid proof of the correctness of an estimate is if it stands the test of the actual dollar payout by the firm that is doing the work. If more is paid out for the work than what was figured on the basis of a given estimating system, even though the takeoff is correct, then the governing job factors have probably not been taken into account.

JOB CONDITIONS

Labor units are based on normal, favorable job conditions. These average units are affected by the particular job factors that govern a given job. The same labor units will not cover work in an area that is clear and work in an area that is obstructed by conveyors, large piping, ducts, or machinery.

This observation applies as well to whether the work is done in low-bay areas or in high-bay areas requiring scaffolds or special lifts. Labor units covering work done on the floors of a highrise building will be affected by the hours consumed waiting for elevators to hoist workers and materials. A congested site with acute problems relating to material storage and handling will affect the labor units. The same is true of work done under winter conditions.

The labor units are based on the assumption that the general contractor is an experienced manager, the plans and specifications are satisfactory, experienced supervision is available for the project, and qualified journeymen are available for the work. Obviously the same labor units cannot possibly apply to all the possible permutations and combinations of job conditions. An estimator will use a standard set of labor units to arrive at a total number of estimated worker-hours to do the work under favorable average conditions, and then factor this total to take into account the particular job conditions that will actually govern the work.

Estimating is not an exact science because the method of foretelling the future in an exact manner has not yet been discovered. However, it can be quite effective and accurate if it is done in a consistent, logical, step-by-step, systematic manner, in which all the items, the obvious as well as the hidden, are taken into account.

Before beginning the takeoff, familiarize yourself with the type of project and review the covering documentation. Study and list all the pertinent notes that appear on drawings and legends and in the specifications. Make sure that you account for any special and unusual work, work done by others, work for other trades, and changes covered by bulletins and addenda. It is important to prepare a binder, divided into systematic sections, where this information can be placed.

Accuracy and consistency are the keys to estimating. Prepare every estimate in the same step-by-step manner, and place the information in designated sections of the estimate binder for easy reference and accountability. Using a binder will ensure that the paper is tied down and you will know exactly where to find the information.

TAKEOFF CATEGORIES

Basically there are three categories of activities in the takeoff process.

Category 1: Items That Are Listed

This involves the major equipment. The major equipment is usually quoted by suppliers in packages covering switchgear, transformers, bus duct, panels, motor control centers, etc. The listing is usually done so as to facilitate the comparison of the various packages quoted by different suppliers. You must be aware of the weight and dimensions of the equipment, and the distance and height that they must be moved for access into the building. You must also cover any special testing requirements and spare parts.

Category 2: Items That Are Counted

This involves counting off items such as outlet boxes, devices, fixtures, and motor connections. Take into account what the fastening requirements are and whether fixtures are recessed, surface-mounted, or hung on chain or rod, singly or in tandem. There are standard columnar estimating sheets for listing the counted items in a systematic manner.

Category 3: Items That Are Measured

This involves measuring the conduit runs, tray, wire, cable, etc. There are standard feeder takeoff sheets and other estimating sheets for listing the measured items in a systematic manner. You must be aware of whether the conduit is in slab or exposed, or suspended on rod, channel, or angle iron and whether it is in a hung ceiling or a block wall. You must also know at what height the conduit will be run and whether the runs can be grouped. Take into account the type of cable, number of conductors, and types of connectors and terminations. Check very carefully that you are using the correct scale, and take into account the possible interference, cutting, and waste that you could be faced with.

Takeoff categories 1 and 2 probably cover 75 percent of the cost of the materials in the estimate, whereas category 3 material may average out to 25 percent of the total material cost. However, category 3 accounts for at least 50 percent of the labor worker-hours, and it is in these operations that the major worker-hour losses usually occur. You must take into account the method of fastening and setting-up time. These can add substantially to the cost of category 3 work.

Very generally, the worker-hours in the average estimate break down as follows:

Supervision	15 percent
Conduit and wire	50 percent
Balance	35 percent

One-third of the worker-hours in the last item (balance) probably pertains to the installation of fixtures. The second item (conduit and wire) is a general designation which takes in all raceways, trays, and cables. Obviously, the above percentages vary depending on the type of job. An accurate labor figure for a given operation can only be obtained by a step-by-step takeoff to which appropriate labor units are applied and factored.

The purpose here is not to describe the step-by-step process of takeoff. There are many excellent manuals and industry data sources that cover this subject in detail. The purpose here is to review the essence of the estimating operation and to point out the strength and the limitations of the procedure, so that you will use it with thought and judgment —not as a routine mathematical exercise that somehow will come up with the true cost.

ESTIMATING FACTORS

You have no doubt noticed that different estimators may use different takeoff procedures and labor units, yet still arrive at a similar cost to do the work. This happens because they account for the various factors in different ways. As long as all the applicable factors are accounted for, in one way or another, the results should be similar. Some estimators will list every locknut and bushing. Others will use composite units. When they apply these to measured or counted items, they will automatically factor in average quantities of the accessory and fastening items that will be required to do the work. Many estimators price the materials using trade prices, so that the applicable discounts will cover cutting, breakage, losses, small errors, and possible price increases. Miscellaneous materials, such as crayons, pencils, flashlight batteries, rags, oil, and soap, are usually covered as a small percentage of the total material cost. No matter how you do your estimating, as long as you know and have tested the limitations of your estimating system, you will have figured out the types of adjustments that you must make to it in order to cover your eventual real costs.

Most labor units are built up to include worker-hours or minutes for the following activities which form part of the unit:

1. Supervision of the crew foreman, pro rata
2. Normal indirect labor activities (activities 1, 2, 5, 6, and 8, described in Chap. 2)
3. Normal material-handling activities (activities 3 and 4)
4. Normal direct labor activity (activity 7), based on good working conditions

The worker-hours consumed by the project staff above crew foreman are usually covered in the worker-hour multiplier, which will be discussed in Chap. 4. Some estimators prefer to cover job staff supervision, and also activity 9 (break periods), as part of job expense. It is not important where these items are covered as long as they are included. Provided that your estimating system is accurate and consistent and you know what is covered where, you are in a position to make the necessary adjustments when the installation conditions differ from your norms.

There are three basic types of factors affecting your work which must be taken into account in your estimate. They are site, contract, and personnel factors, a list of which follows:

1. Site factors
   a. Type of project
      (1) Typical or nonstandard
      (2) Commercial, industrial, or government
      (3) Distance to work areas
      (4) Congested or clear
   b. Height
      (1) High-bay or low-bay
      (2) Highrise building
   c. Project location
      (1) In town or out of town
      (2) Accessibility
   d. Material handling and storage facilities
      (1) Availability of loading ramps and hoisting facilities
      (2) Types of storage facilities

e. Security

(1) Danger of flooding

(2) Fire hazards

(3) Safe storage

2. Contract Factors

a. Quality of design

(1) Completeness of drawings and specifications

(2) Coordination of layout and interfacing

(3) Reputation and experience of professionals

b. Quality of general contractor or construction manager

(1) Management reputation

(2) Availability of experienced staff

(3) Experience in this type of work

(4) Busy or not with other projects

c. Type of schedule

(1) Normal, or will the work be dragged out?

(2) Accelerated, requiring overtime or shift work

(3) Subject to delays, disputes, interruptions

d. Beneficial use and occupancy

(1) Right of owner to make use of facilities before contract is completed

(2) Use of electrical system for temporary lighting and power

3. Personnel factors

a. Availability of skilled personnel

b. Ability to work with optimum crew size

c. Past experience in this type of project

d. Typical or atypical work

e. Weather conditions

f. Height and congestion conditions

g. Availability of adequate hoisting facilities for crews and materials

h. Morale and psychological problems

You, and the design professionals who are checking your estimates, should account for these factors in order to establish the amount of labor that will actually be required for the work. If possible, try to negotiate a percentage factor to be applied to your estimated labor which is derived from your standard labor units.

There are estimating systems and industry data sources that quantify the impact of these job factors. You tick off the applicable job factors and add up the resulting percentage, which you use to adjust the estimated labor. There are other estimating systems, having up to five columns of estimating units to suit the various degrees of difficulty to which the work will be subjected. No doubt you have your own system. It is important to involve your field people in the estimating process, so they will understand that this is a dynamic living process requiring the feedback of their field experience to ensure its accuracy.

Keep in mind, as well, that the most accurate takeoff can be ruined by sloppy material pricing. Mistakes are made because you have not ascertained whether the material is priced per hundred or per thousand, whether the tax is included or not, or whether the pricing catalog is out of date. Identify the material properly, so that you will get accurate prices from your suppliers. Protect yourself against possible price increases. Make sure that you add your hidden costs and job expenses. These cover items such as subcontract work, cartage and delivery, cutting and patching, painting, temporary wiring, field office and storage, bonding, insurance, inspection, drawings and drafting expense, traveling expense, contingency, and guarantee. An ever-increasing hidden cost is financing. Slow cash flow and holdback are very expensive to finance and put great stress on your operation.

You now have the prime cost of the estimate, to which you must add your overhead and profit to arrive at the total cost. (This process will be covered in greater detail in Chap. 4.) Take the time to explain to your field people, and to the professionals with whom you deal on your jobs, the basic principles and procedures of your estimating system and the way in which you prepare your estimates for tender and for change-order quotations. This ongoing dialogue, combined with the regular feedback of job experience, will help you double-check and fine-tune the labor units and job factors that form the basis of your estimating system.

ESTIMATING TECHNIQUES

Assuming that the estimate has been prepared with care and accuracy, the individual items very often have a tendency to complement or compensate for each other. Some may be a little high, others a little low. They tend to average out. It's much like a choir: If you single out any given singer, he or she may not be particularly good, but as a whole the choir may be very good. This must be understood, particularly by the professionals who review or check an estimate. They must understand that it is the result as a whole, and not any individual items, that must add up to a value covering the actual labor and material that will be consumed when the work is eventually done.

A contractor will be successful on only a fraction of the jobs on which he or she bids. Estimators are therefore confronted with the need to develop techniques that will speed up the estimating process without decreasing the accuracy to unacceptable levels. Square-foot prices based on previous jobs can be used for budgeting or double-checking purposes, but never as a legitimate estimating system. No two jobs are exactly alike, and to obtain an accurate cost you must do your takeoff step by step. The use of composite units for assemblies is an efficient and acceptable method of estimating. By measuring and counting the main items, such as outlet boxes, conduit, wire, and fixtures, the composite unit will pick up the accessory materials and fastenings.

ESTIMATING WITH COMPUTERS

Although computers and computer estimating programs are readily available, they have not, as yet, made significant inroads in the construction industry. In an estimate, about 75 percent of the takeoff represents only about 25 percent of the material cost. The cost of material is subject to frequent changes. Most computer estimating systems deal mainly with the measuring and some of the counting activities of the estimate—that is, with the 25 percent material-cost portion of the estimate.

Computer estimating is very often based on elaborate coding systems. The estimators must still go through the step-by-step procedure of measuring and counting. What they do save is the listing and pricing of the standard material and labor—and even here they still have to review the material and labor, to take into account price increases and labor factors. They also have to write in the listed materials and the special counted material, such as the different fixture types, and other specialized items of this nature. Because contractors must estimate many jobs before they are successful on one, there will have to be a trade-off between making computer estimating much simpler to use and still maintaining an acceptable level of accuracy. By trying to make the system do all things—for example, making the codes sufficiently detailed to identify the exact size, depth, mounting arrangement, and number of knockouts of a box—the computer estimating systems have become too complicated.

The function of an estimate is to come up with an accurate cost of doing the work in the shortest time and the most efficient manner possible. When you get the job, the next step is to fine-tune the estimate so that general categories of accessory materials are converted to exact listings, based on your job coordination and installation requirements. There is no doubt that computer estimating will eventually be tailored to suit your requirements and will become more prevalent in the years to come.

ROLE OF THE ESTIMATE

You and your associations must campaign continuously to involve contractors, their field people, and the industry professionals, the architects and engineers, in educational programs to improve their real comprehension of the art and science of estimating. An informed competitor is a good competitor, and an informed engineer will be fair and reasonable when called upon to establish the value of a package of work. Project managers and foremen must know not only how to install the work, but also how much it costs to do so and how many worker-hours it should take.

What the field must therefore know about estimating is that the estimate is the source of reference from which they will obtain the information to establish the worker-hour targets for each and every operation that forms part of the project. From this source information, they can establish the size of a crew and the number of working days for this crew to complete a given operation. With this source information and the project schedule they can determine the maximum size of the gang based on the quantity of material that is planned for installation in a given time.

The estimate is a detailed picture of the job and the basic tool for planning and controlling it. Estimating teaches you that there is a definite relationship between the quantity of materials installed and the number of worker-hours that the installation should take based on the prevailing job conditions. This is the lesson that the field must learn from estimating.

4
Costs Are Higher Than You Think

The purchasers of your services generally believe that contractors charge too much for their work. This is expressed most vociferously when the time comes for negotiating the charges for extras to a contract.

The fact is that too many of you have only an incomplete understanding of the real cost of doing electrical work. You are too often on the defensive when explaining this vital aspect of your business. You personally, and your associations, have done a poor public-relations job in educating owners, engineers, and architects as to the true nature of your costs. The myth that material cost plus direct labor cost adds up to total costs and that the old ten-and-ten markup covers everything else does not stand up particularly in these changing times. The jobs are getting bigger, more complicated, and more risky. They are subject to the vagaries of the high-cost-material and labor markets. They require more planning, management, and coordination than ever before.

Every contractor understands thoroughly that a bid constitutes nothing more certain than an estimate. An estimate is the application of a formula, in order to arrive as close as possible to the reality of the final cost for work that will be done in the future, under conditions that will not be fully known until the actual work is done. The perfect system for accomplishing this has not been devised. The possibility of error, the guesswork, and the risk cannot be completely eliminated, but they can be reduced if the contractor understands and takes into account all the factors that go into making up the cost.

This chapter will attempt to identify and analyze the specific areas of cost in the operation of an electrical construction business, and to offer a formula for providing for them in the estimate. The figures and percentages shown may not apply exactly as listed for every project or every operation. These figures and percentages are historical averages and are to be used as a guide. The fact is that the costs they represent exist and must be accounted for, if you are to remain in business.

FOUR MAIN DIVISIONS OF COST

There are four main divisions of cost associated with every electrical installation:

1. Material cost, including all applicable taxes
2. Labor cost, including labor burden
3. Direct job expense
4. General overhead

Of these four items, the most controversial and least understood is direct job expense. Most people confuse direct job expense with general overhead, but these are two separate areas of cost. *General overhead* pertains to items of expense that cannot be identified with any particular project. On the other hand, *direct job expenses,* as listed in Table 4.1, are generated by a particular project and are as much a direct cost as the material and labor that go into it.

DIRECT JOB EXPENSE

Direct job expenses are generated in the two ways shown in Table 4.1:

1. As expenses incurred in the management of labor
2. As expenses incurred in the management of materials

In some estimating systems, items 1 to 5 of direct job expense are covered by a labor factor which is applied to the estimated direct worker-hours. In this analysis, these items of job expense are included in the labor multiplier. This allows the multiplier to be checked in the field by dividing the total job payroll by the actual field worker-hours consumed for any given period.

Items 6 to 14 can be accounted for as a separate percentage applied to labor and material. Usually, though, one average percentage is applied to the total of material plus labor to account for this item of cost. In one way or another, these items of direct job expense must be included in the cost of doing the work.

The confusion surrounding this category of cost arises because job

**TABLE 4.1   DIRECT JOB EXPENSE APPORTIONED
BETWEEN LABOR AND MATERIAL**

JOB EXPENSE	LABOR	MATERIAL
1. Project manager and general foreman	Items 1 to 5 range from 10 to 30% of direct labor cost and are included in the labor multiplier	
2. Coordination and drafting		
3. Clerical and timekeeper		
4. Storekeeper		
5. Shop steward		
6. Permit and inspection fees	Items 6 to 14, as they pertain to labor, range from 5 to 10% of direct labor cost, and are accounted for as a percentage of the labor	Items 6 to 14, as they pertain to material, range from 5 to 10% of material cost, and are accounted for as a percentage of material
7. Tools consumed and depreciated		
8. Freight, carriage, and delivery		
9. Stationery and supplies		
10. Job shack and storage		
11. Telephone and telegrams		
12. Job visits, room and board		
13. Contingency and guarantee		
14. Bonding and financing		

expense is mistakenly thought to be part of general overhead. This fallacy must be cleared up, not only in the thinking of some customers, but also in the thinking of some contractors. Job expense is a direct job cost in the same way as job labor and job material are direct job costs.

As construction becomes more and more complex and sophisticated, increasing staff must be assigned to every project to interface the work with other trades, to plan, to coordinate, to constantly review the drawings and specifications, and to account for the inevitable revisions and changing job conditions. Interference drawings, field sketches, and as-built drawings must be prepared. Diagrammatic and specification details must be translated into working drawings. Depending on the magnitude of the project, coordinators, engineers, and draftsmen are assigned to the project for this purpose and are a direct job expense.

It is the responsibility of the project staff to convert the design into a completed installation. To complete the installation they must overcome the constraints of the actual job demands and conditions. Revisions and change orders must be checked, estimated, negotiated, and incorporated into the work. Job staff is required to carry out these estimating, coordinating, and planning functions. Staff above the field-foreman level must be assigned to lay out the work, schedule it, manage it, and control it.

Material handling is an item of cost, the implications of which you do not fully acknowledge or account for. Table 4.1 only accounts for the cost of the storekeeper and the storekeeper's assistants. A detailed study of any job would show that, on the average, 20 percent of the job worker-hours are expended in receiving, storing, and delivering material to the working crews, as well as in searching out and moving around the materials that go into the work. How many times are productive operations halted, so that the workers can be utilized to unload a van of cable reels or conduit? Labor units include allowance for a normal amount of material handling. Excessive material handling must be factored into the direct worker-hours required to do the work.

The cost of items 6 to 14 of job expense can be derived from the contractor's cost records with varying degrees of ease. Some job-expense items, such as permit and inspection fees, and bonding and financing, can easily be established as a percentage of material and labor cost. The cost of the other expense items, such as tools consumed and depreciated, and contingency and guarantee, are more difficult to establish, but they must be worked out and accounted for.

Contingency and guarantee are job-expense items that fall into two categories. First, since your workers are human, you cannot expect them to perform with absolute perfection. Thus you are obliged to make good that portion of their workmanship that is faulty or incorrect. Similarly, a small percentage of the material which you receive from suppliers is incorrect or faulty. A supplier will usually make good on such items, but in most cases you must pay for the labor to remove and replace the faulty material. Thus, for every worker-hour of work, you must provide a percentage to cover such contingency and guarantee. Second, you must replace stolen material or material damaged by your work force or others through accident or negligence. Thus, for every dollar of material which you purchase, you must also provide a percentage to cover contingency and guarantee.

Direct job expense, which is covered partly in the labor multiplier and partly as a percentage of material plus labor, must be recognized, costed, and accounted for as a legitimate cost. Direct job expense added to the cost of material plus labor forms the direct cost of doing the work. To this direct cost you must add the cost of the *general overhead* in order to arrive at the total cost.

GENERAL OVERHEAD

General overhead exists independently of any particular project. It consists of items of general expense that your company must expend in order to operate as a business. All the jobs that your company has in progress at any given time must collectively generate sufficient contribution to cover the cost of this general overhead.

A contractor must maintain an estimating and often an engineering staff. The missionary work done by this staff on jobs tendered but lost must be covered. There are many times when you must tender ten jobs to get one. There are expenses involved in the operation of your office and your central warehouse, including the cost of salaries of their respective personnel. Add to these the expenses for rent, light, heat, telephone, stationery, advertising, financing, legal fees, and automobiles and trucks, and you arrive at a cost of general overhead which can average 15 percent of the direct costs as previously detailed.

Table 4.2 lists the items that together form the general overhead. They are grouped under the general headings of administrative salaries, office salaries, staff salaries, operating expense, fees, and financing, all of which must be covered as part of operating your business. It is important that you are aware of the cost of this general overhead and that you account for it in your pricing. Not understanding the real cost of direct job expense and general overhead has forced many contractors into bankruptcy. These costs are as real as the obvious costs for labor and material.

Since labor is common to all jobs, whereas the cost of materials may vary depending upon whether the material is supplied partially or completely by the contractors or by others, many contractors figure their direct job expense and general overhead as a percentage of the labor cost. Whatever the manner used to calculate these items, they must be figured and accounted for if you are to remain in business.

ESTIMATING THE REAL COST OF THE WORK

Job costs are like an iceberg: What is obvious above the surface is only part of the total cost; the remainder lies hidden and must be identified so that it will be accounted for. In every estimate, the quantities and costs of various materials are calculable, and thus they form the basis for the formula to establish the total cost. They represent the part of the iceberg that is visible and measurable. Nevertheless, even here you often neglect to include allowance for miscellaneous material items, spoilage, cuttings, and additional lengths required to get around ob-

**TABLE 4.2   ITEMS THAT MAKE
UP GENERAL OVERHEAD, AS
PERCENTAGES OF TOTAL COST**

EXPENSE CATEGORY	PERCENTAGE OF TOTAL COST
Administrative salaries	2.5
Office salaries	3.0
General office	
Accounting	
Bookkeeping	
Costing	
Filing	
Telephone operator	
Staff salaries	3.0
Engineering and estimating	
Purchasing	
Warehouse	
Shop	
Operating expense	3.0
Rent	
Power, telephone	
Office equipment and furniture	
Supplies, stationery, postage	
Autos and trucks	
Sales promotion	
Fees	2.0
Legal and audit	
Advertising	
Insurance	
Business taxes and licenses	
Financing	1.5
Holdbacks	
Bad debts	
Reserve	
General overhead	15.0

structions and fastenings. These must be accounted for, since they will surely be required to do the installation.

To the takeoff list of materials, which you hopefully assume to be complete and correct, you apply labor units to derive the worker-hours required to do the work. These units cover the field labor required, including the field foreman. It is evident that just as one size of hat cannot fit every head, so one set of labor units cannot fit every type of job or job condition. Labor units are historical averages that must be factored to fit a particular job. The factor which is applicable to the labor units must take into account the time lost due to the constraints of working height and job conditions, the caliber of the construction team, and the caliber of the available labor force. Experience and judgment must be applied in factoring the worker-hours derived from labor units. These factors can be field-checked, because lost time and productivity are measurable. The correct labor estimate is that which will cover the actual cost of field labor, assuming that the contractor has done his or her best to plan, manage, coordinate, and control the work.

The essence of the formula contained herein is that, for every dollar expended on material and labor, the job must generate revenue to pay for direct job expense and general overhead. To simplify the procedure and the calculations, it is suggested that direct job expense should be derived as a percentage of material plus labor. The general overhead should be derived as a percentage of material plus labor plus direct job expense.

PRICING CHANGE ORDERS

You are no doubt often faced with a familiar area of controversy: the pricing of change-order quotations. There is a common tendency to confuse the manner of pricing cost-plus work with the manner of pricing change-order quotations.

In cost-plus work there is no risk. All the costs, including all direct job expenses, are known and accounted for. Thus labor in this case can be charged out at payroll rates, plus burden and agreed-upon markups for general overhead and profit.

However, in the case of a change-order quotation, the price is arrived at as an estimate of work shown on a flat plan to be performed at a future date under conditions which cannot be fully foreseen. Any error, or any impact of this change on the existing contract, will appear as an additional cost to the contractor. There is therefore a substantial element of risk in making up the estimate.

As was stated previously, the preparation of an estimate starts with the material takeoff. Many contractors charge for the materials in a change-order quotation at trade prices. They rely on the applicable discounts to cover them for small errors, cuttings, spoilage, price increases, and other additional cost impacts. Labor units are applied to the material takeoff to derive the worker-hours of labor. Every change to the work has to be done under particular prevailing job conditions, and at a point in the cycle of the overall work which may be unfavorable in varying degrees. Since most labor units are based on average favorable conditions, the resulting labor worker-hours must be factored to take these actual job conditions into account. The factor to cover lost time and actual productivity can be checked and verified in the field. You are entitled to charge for your real labor costs and not to accept the labor derived from units that don't reflect the conditions under which the change will be done.

Many items of lost time have developed historically in your industry, and many of them are sanctioned by labor agreements. Table 4.3 itemizes the portion of lost time that is covered by labor agreements. In most cases the lost time itemized in Table 4.3 is much less than what is actually lost on the job. Break periods are limited to 15 minutes in most labor agreements, but in fact they get stretched to 30 minutes or more in the field. You accept this lost time as a fact of life, because it is part of the culture of your industry. You must make strenuous efforts to limit this lost time to the letter and spirit of the agreements, but nonetheless you must recognize that it exists and must be charged for.

Under the best of conditions, as shown in Table 4.3, you lose over two hours of every eight-hour day for which you pay your workers because of break periods sanctioned by labor agreements. This time is lost by even your best workers and is not to be confused with the additional time lost due to material handling or reduced productivity. Productivity and efficiency have to do with the way in which the project is run, with delays and interference, and with the caliber of the labor force and of management. They will be covered in subsequent chapters.

LABOR MULTIPLIER

To establish the full size of the iceberg (that is, the total labor cost), the top of the iceberg (that is, the estimated worker-hours derived from labor units) must first be factored and then converted into dollars by

TABLE 4.3   TIME CONSUMED BY BREAK PERIODS (ACTIVITY 9)

REASON FOR BREAK	TIME, MINUTES
Punch in, change clothes, select tools, and go to point of work	15
Morning break period, including walk to and from shack	30
Preparation for lunch	15
Return to work after lunch	15
Afternoon break period, including walk to and from shack	30
Relief periods	30
Preparation for quitting work and punch-out	15
Total	150

means of a *labor multiplier.* This multiplier is based on the prevailing labor rate plus a percentage to cover the indirect labor cost associated with it. It can be field-checked by dividing the total site payroll by the quantity of field worker-hours covered in that payroll.

The form utilized to work out the multiplier is illustrated in Fig. 4.1. In item 1 of Fig. 4.1, list the hourly rate of an electrician. In item 2, list the percentage that the indirect labor (over and above field foremen) constitutes of the direct labor force. For example, if there are five supervisory personnel for a field crew of 25 which includes the field

Labor Multiplier

Job name _____   Job no. _____

_____   Date _____

Multiplier will be effective until _____

1. Basic rate including labor burden $_____
2. Number of indirect workers
    Project manager   _____
    General foreman   _____
    Coordinator   _____
    Office clerk   _____
    Storekeeper   _____
    Runner   _____

    Total indirect workers [ ]
    _____ = _____ % $_____
    Total direct workers [ ]   (of item 1)
    including field foremen

3. Union steward   _____ % $_____
                                            (of item 1)
4. Job factor   _____ % $_____
                                            (of item 1)
5. _____   _____   _____

    Total $_____
    Labor burden   _____ % $_____

    Multiplier $_____

FIGURE 4.1   Form for deriving the labor multiplier.

foremen, then this percentage will be 5/25 or 20 percent. Twenty percent of the electrician's rate will be listed in item 2. In item 3, list the percentage pro rata that the union steward constitutes of the direct labor force. In the case given it works out as 1/25, or 4 percent. By adding the above items you obtain the multiplier for converting worker-hours to dollars. As was indicated previously, the multiplier can eventually be checked in the field by dividing the total payroll by the quantity of field worker-hours covered by it. To the total estimated labor cost you have to add the labor burden, which consists of items shown in Table 4.4. The labor burden varies from area to area and is based on particular local labor agreements and laws.

TABLE 4.4   LABOR BURDEN PERCENTAGE APPLIED TO THE LABOR COST

LABOR BURDEN ITEM	PERCENTAGE APPLIED IN 1982
Vacation pay	
Joint committee	
Health insurance	
Pension plan	
Worker's compensation	~30
Unemployment insurance	
Social security and indemnity fund	
Public liability and public damage insurance	

To the total of material cost plus labor cost you then have to add a percentage to cover the direct job expense, as shown in Table 4.5. This gives you the total direct cost to which you apply your overhead and profit.

In the final analysis, an estimate for a given amount of work is the result of a formula for arriving at a cost that will equal the actual cost of the work when it is done. The formula can only achieve this result if it takes into account the hidden as well as the obvious costs. The formula can succeed only if it is based on field knowledge, experience, and judgment.

TABLE 4.5   JOB EXPENSE

EXPENSE ITEM	SIMPLE JOBS		COMPLEX JOBS	
	MATERIAL	LABOR	MATERIAL	LABOR
Permits and inspection fees		3.0		3.0
Tools (consumed and depreciated, rental, supply)	2.0	3.0	5.0	5.0
Freight and cartage	0.5	0.5	2.0	1.0
Telephone	0.25	0.25	1.5	0.25
Field office and storage	0.25	0.25	1.5	0.75
Total	3.0	7.0	10.0	10.0
Average percentage, to be applied to the total of material plus labor		5.0		10.0

5
Change Orders

Changing or adding to the work of a contract that is in progress is expensive, usually not because of the contractor's overcharge for the additional work (commonly called *extras*), but because of the disruptive effect that changes have on the work in progress and because of the unfavorable job conditions under which the changes may have to be made. The emphasis should not be on reducing the change-order quotations to levels at which the contractor may not recover his or her real cost; the more reasonable approach would be to reduce the quantity of unnecessary changes by more careful and responsible preparation of the plans and specifications in the first place. (An ounce of prevention is worth a pound of cure.)

When changes are called for, the contractor should not be put in the difficult position of having to defend an estimate against the owner's immediate assumption that it is excessive. Too many people believe that a contractor automatically makes money on extras. No doubt many contractors believe this to be a fact, as well. So much time is spent arguing about the cost of extras, and so much ill feeling is generated as a result, that the subject warrants your study and analysis.

If you have ever sat in front of an open-hearth fireplace and were warmed by the intense heat of its roaring fire, you may have thought that it is a very efficient way to heat a room. Yet studies have shown that a fireplace is very inefficient. Heat goes up the chimney and already-warmed room air is used up by combustion. The result is that you are getting a lot less net heat than you might think, and in some cases there may even be a net loss of room heat. Illusions die hard, however, and people still love fireplaces. This analogy applies to the effect that changes have on a contract.

IMPACT OF CHANGES

When contractors tender on a job, they obtain a set of bid documents consisting of plans, specifications, and instructions. A bid is based on the information contained therein. If it is successful, this information is used to plan and schedule the contract. At time of tender a contractor does not have a crystal ball to foretell the future. The competitive tendering system forces contractors to figure on the information contained in the tender documents in hand. They have no way of knowing if there will be changes, when they will happen, what their magnitude will be, or how they might affect the contract should the bid be successful. Contractors cannot estimate the impact of possible changes which are unknown, nor are they required to do so contractually. Usually there are provisions in the tender documents and in the contract on the manner of dealing with change orders.

A large volume of changes, which increase the original contract by 10 percent or more, will have an impact on the cost of performing the original contract, and in some extreme cases can change the very scope and nature of the contract. The contractor is warmed by the prospect of making some money on the extras, but is usually not aware of the impact costs and the resulting losses going up the chimney.

The antagonisms that very often characterize the relationship between the contractor and the party who is checking quotations or claims for changes to the contract stem from the following facts.

CORRECTIVE CHANGES

About half of the changes on a job are *corrective changes.* Their purpose is to correct errors in design and to eliminate areas of uncertainty in the specifications. Specifications are often assembled from parts of other jobs and may not be exactly tailored for the project in hand. There may have been insufficient communication between the owner and the designers. There may have been insufficient coordination, integration, and interfacing between the designers of the various specialties that together make up the total design package.

Corrective changes must therefore be made so that the project can be properly built and serious interference between the trades can be eliminated. Human nature being what it is, corrective changes, which cost money but add very little to the value of the project, can lead to buck-passing and arguments.

BENEFICIAL CHANGES

The other half of the changes on a job are *beneficial changes.* The owner and the designers may decide to add new energy-saving systems to the project. There may be tenant requirements or new products and technology that they have decided to incorporate into the design. All of these add beneficial value to the project.

However, changes are expensive, and the expense varies, depending on the size of the changes, at what point in the project they are made, and how they affect the overall job schedule. A small change can be easily worked in, but a large quantity of small changes will result in disruptions of job schedules and work-force deployment. If the completion date is critical and cannot be extended, then the change may force the contractor into overtime or overstaffing the job, both of which are costly and losing propositions. In any event, extensive changes will disrupt the planned use of supervisory personnel, tools, and crews and will add greatly to the cost of performing the overall work. Beneficial changes are easier to deal with than corrective changes, but there are always arguments about the cost.

IMPACT CHANGES

There are also *impact changes.* These are very rarely treated as changes, and the contractor must at the least act to control and reduce their adverse effects. Mostly they are paid for, if at all, as part of a hard-fought claim.

Impact changes result from changes in the schedule due to delays in building the project, late or obstructed availability of the working areas, interference by other trades, holds placed on portions of the work, changes in job conditions, and too many changes of all kinds. The topic will be covered in subsequent chapters. If an impact change is treated as a change at all, it will fall into the category of corrective changes, which add little to the beneficial value of the project but much to the volume of disagreements and arguments that occur on many jobs.

CHANGED CONDITIONS

The market value or the beneficial value of a change to the owner is usually much lower than the actual cost to produce the change under the particular job conditions that affect this work. The change may require you to dismantle work that has already been installed. This is costly and also very demoralizing to the workers. Material already on the site, which must be returned to the supplier because of the change, is subject to restocking charges. The revised material may be subject to long delivery which may disrupt your work schedule. The purchasing agent usually negotiates the best prices on packages of materials or equipment. Any subsequent changes to these packages, whether additions or deletions, will interfere not only with the orderly production of shop drawings and delivery, but also with the pricing, which will not be particularly favorable. To absorb a change, particularly a substantial change, into the work in progress is disruptive and expensive.

When changes are made, the following chain of events is set in motion and consumes worker-hours:

1. The project manager advises the foreman of the change. The foreman's planned schedule is disrupted.
2. The foreman reviews the work, examines the location, marks out the work, and orders the material.
3. Material and tools are removed from the old work, and the new material and tools required are moved in for the new work.
4. The foreman removes workers from an existing activity and moves them to the new work.
5. When the change is completed, the workers are shifted back to the original activity.
6. Expenses are incurred in estimating the change and in integrating the change into the work schedule.

An increase in cost will result from changed conditions, even if the type of work being done is unchanged. If changes are estimated on the same basis as the original contract, you will be faced with a deficit rather than a profit. You must anticipate the above factors when pricing a change order. Your labor costs can increase drastically because of the resulting adverse conditions, particularly if your planning and scheduling are badly disrupted. Changes, like inflation, will have the effect of requiring more supervision and more labor for the same amount of work, thus bringing up the cost.

RECOVERY OF REAL COST OF A CHANGE

The cost impact that the change may have on the original contract is almost never accepted as a legitimate factor to be taken into account when arriving at an acceptable value for the change. This problem is compounded by the difficulty of visualizing or quantifying what the impact cost really is or will be. Particularly with a large quantity of small changes, it is difficult to determine the impact effect of each change, even though the total effect may turn out to be quite damaging.

The professionals who check the extras rarely have personnel with the required field experience and estimating experience to be able to cost or evaluate the work in the context of the real job conditions. The prevailing procedure is to require the contractor to prepare a detailed estimate of the change. The checker then feels bound, in proper performance of duty, to tear the estimate apart and cut it down. When owners and their professionals initiate changes, they should be responsible for the consequential effects. They, after all, are the ones making the changes, which in the context of the construction process today must turn out to be relatively expensive. It is up to them to control the quantity of changes, particularly corrective changes, by better planning and control in the design stage. They should not expect the contractor to make good on their mistakes, omissions, or revisions as part of the contract.

Some contractors, when pricing an extra which obviously will be done at some time in the future under conditions difficult to foresee, will build some padding into the price to protect against contingencies and price increases. Probably without realizing it, they are also building in some protection against impact costs. In the absence of a more businesslike understanding of their real cost, it may very well be this padding that is keeping them in business. The checker, on the other hand, will try to remove the padding and cut down the extra to the bare minimum.

What should be aimed for by all parties concerned is a sound estimating procedure for change-order work, which will identify the applicable impact factors and enable the contractor to recover the real cost. In one way or another a contractor will attempt to recover the real cost of doing the work, in order to remain in business. Much needs to be done to establish an acceptable, fair, and reasonable procedure for pricing extras in the construction industry.

Most contractors are in favor of standardizing a procedure for pricing extras that would enable them to recover their costs, yet be fair to the owners and expedite approvals. Change-order quotations, however, must reflect the proven fact that it is more expensive to do the work called for in a change than to do similar work as part of the original contract. If a change is made before the work in the given area has been started, then the additional expense generated by the change will be at the low end of the scale. If the same change is made when the work is in progress, then it is doubtful that even substantiated factoring will cover its real cost, and the additional expense will be at the high end of the scale. In any event, rapid approval of changes is necessary to mitigate the impact which the change will have on the work in progress.

Work on a job is like an assembly line. It takes time to learn and absorb the details and requirements of the plans and specifications and to find out where things are, where things go, where work is to be done, and by whom it is to be done. This is called the *learning-curve effect:* time and effort are required for workers to come up to speed and develop an acceptable rhythm and tempo of work all along the line. Changes will result in breaks in the line, and these will affect the rhythm not only of the revised work, but also of work all along the line.

You can imagine the devastating effect that delays will have if vital operations are held up for long periods while change-order approvals are waited for. Studies have shown that 50 percent of a job cost overrun is due to factors which reduce labor productivity. Changes that must inevitably affect the tempo and rhythm of the work, and late approvals, are an important contributing factor to the reduction of productivity.

Owners and their professionals are sometimes upset when contractors do not mark up a change that involves a reduction in the work (commonly called a *credit*). Credits are not marked up because they disrupt the smooth running of a job in the same way that an extra does. The overhead portion of the credit has been used up in the original coordination and planning of the work and in the disruptions involved in removing the work. Owners and their professionals must realize that by making changes they are disrupting the work and increasing its cost because of the reduction in productivity that results.

CASE HISTORY

As an example, Contractor E was awarded a contract in the amount of $1.1 million to do the following work:

1. Supply and install embedded conduits.
2. Supply and install underground ducts, including excavation and backfill.
3. Install only the 12.5-kV switchgear supplied by the owner, including all necessary wiring and connections.
4. Supply and install the power transformer.

There was less than 1 percent difference between the low bid and the second bid, and between the lowest and highest bidder there was only a 7 percent spread. The bidding was very close, and it can be assumed that the contract price was adequate to do the work called for. The contracted completion period was seven months. The actual completion date was 20 months after the award of the contract. The delay to a large extent was due to the numerous changes that were made. In all, there were 62 change orders amounting to $450,000, which worked out to 41 percent of the original contract.

Contractor E suffered a 32 percent overrun of worker-hours over and above the total of estimated contract worker-hours, plus the worker-hours for the extras. You would think that the large percentage of extras on this project would have automatically guaranteed the contractor a good profit on this job. In fact, his markup was consumed by the substantial worker-hour overrun, which to a large extent can be attributed to the frequency and magnitude of the changes, the long delay in obtaining approvals, and the disruptive impact effect these had on the performance of the work.

6
Your Contract and Impact Costs

For years contractors have accepted the impact costs of many job factors which are beyond their control and which are lumped under the general heading of job conditions. Construction is such a dynamic and fast-changing industry, in which no two jobs are ever exactly alike, that you and your field personnel have come to accept these impacts on the cost of doing the work as a fact of life.

IMPACT COSTS DUE TO DELAYS AND DISRUPTIONS

Impact costs flow mainly from delays and disruptions of the normal contract schedule. The result very often is a labor cost overrun. There are two types of delays that have an impact on labor costs.

First, delays result from design problems because

1. Drawings are poor and require correction and clarification. The design is ambiguous.
2. There are excessive changes.
3. Decisions are slow in coming, and responses to questions seeking design clarifications are late.
4. Approvals are late in coming.
5. Parts of the work are put on hold.
6. There are design errors and omissions.
7. There are installation difficulties because of space problems or because specified items are not readily available.

Second, delays result from coordination and schedule problems because

1. Coordination by the general contractor, or the construction manager, or the owner, is inadequate.
2. Coordination of the work of the different trades is inadequate.
3. There is interference by the work of other trades.
4. Areas of work are not available on time.
5. Equipment supplied by others is not available on time or is delivered at the wrong time.
6. The project falls behind schedule because of lack of performance by the general contractor, the construction manager, or the owner.
7. Adverse site conditions exist, or adverse weather or strikes occur.
8. The work suffers due to acceleration or extended duration.

The additional costs flowing from these delays and disruptions can be extensive and sometimes catastrophic. The momentum and productive rhythm of the work are broken. Many operations are slowed or stopped, and crews must be shifted to other locations. The orderly flow of information and materials is disrupted. The job coordination and planning must be revised over and over again. Delayed work may have to be done later under adverse conditions. Payments for work are delayed, with the result of additional financing costs. Job earnings, which have been budgeted to cover the overhead for the contract period, must cover a longer period when the contract is extended, and, as a result, revenue is lost.

The result of delays to the work is that productivity decreases and the cost of supervision increases. The rate of performance will decrease while the installers mark time waiting for the correct or missing information or for materials to catch up the work. Field studies show that there is a worker-hour loss every time workers are shifted from one operation to another, and every time they are added to or laid off from a working crew. When operations are disrupted because of delays, poor planning, faulty coordination, lack of information, or changes and workers must as a result be shifted to other operations or laid off, each such movement involves some or all of the following lost-time activities:

1. Sort out and record the material and tools.
2. Return material and tools to the stores.
3. Move to the new location.
4. Review the new work.
5. Order material and tools for this new work.

6. Receive these materials and tools and set up.

7. Get familiar with the new location and the new work.

These activities can consume two to four hours per worker per move. Some shifting around is inherent in the nature of construction or can be attributed to normal job conditions. However, when this shifting around is caused by factors beyond your control, then this lost time must be recorded for a possible claim. Delays invariably result in a loss of productivity and an overrun of worker-hours.

CASE HISTORY

Take the following actual case as an example. Electrical Contractor E was awarded a contract in the amount of $5 million for the electrical work in a large water filtration plant. The project had a contracted completion time of 22 months. As it turned out, the project was actually completed in 38 months, which represented a gross overrun of 16 months.

The electrical contract was a unit-price contract. Quantities were listed for every unit, and the units were broken down into a material component and a labor component. Progress billing was based on actual field measurements of work installed during the given month. It was possible, therefore, with a high degree of accuracy to compare the worker-hours earned during the month with the actual payroll worker-hours consumed during that same period. Thus there was a continuous comparison between the worker-hours actually spent and the estimated worker-hours earned for the same work during any given month.

This comparison is tabulated in Table 6.1. You can see by studying this table that at the completion of the project, the actual worker-hours consumed, on the basis of payroll records, were 33,000 more than what was estimated. This works out to an overrun of 15 percent.

The specifications called for designated portions of the project to be completed on given milestone dates. The building portion of the project had been proceeding for some time before Contractor E was awarded the electrical contract. Upon arrival on site he realized that the project was way behind schedule and that there was no way that the milestone dates would be met. In fact, Table 6.1 shows that very little electrical installation work took place for the first eight months of the contract. After 16 months, 56,000 worker-hours had been spent, according to payroll records, to earn 58,000 worker-hours according to the estimate. Up to this point Contractor E was well within his labor

TABLE 6.1   COMPARISON BETWEEN WORKER-HOURS EARNED (BASED ON FIELD MEASUREMENTS OF WORK INSTALLED) AND ACTUAL WORKER-HOURS (BASED ON PAYROLL DATA)

MONTH	HOURS EARNED MONTHLY	HOURS EARNED CUMULATIVE	HOURS WORKED MONTHLY	HOURS WORKED CUMULATIVE
1	....	....	164	164
2	....	....	136	300
3	....	....	292	592
4	....	....	154	746
5	....	....	478	1,224
6	....	....	849	2,073
7	639	639	1,317	3,390
8	107	746	2,428	5,818
9	4,115	4,861	2,864	8,682
10	3,913	8,774	4,664	13,346
11	5,592	14,366	5,553	18,899
12	10,687	25,053	7,860	26,759
13	11,046	36,099	8,372	35,131
14	13,971	50,070	7,491	42,622
15	7,623	57,693	8,037	50,659
16	425	58,118	5,360	56,019
17	....	58,118	4,271	60,290
18	....	58,118	565	60,855
19	10,072	68,190	2,235	63,090
20	8,021	76,211	15,600	78,690
21	11,130	87,341	15,331	94,021
22	10,640	97,981	13,605	107,626
23	15,851	113,832	14,257	121,883
24	13,119	126,951	14,605	136,488
25	13,595	140,546	17,608	154,096
26	6,052	146,598	14,807	168,903
27	7,529	154,127	14,576	183,479
28	7,647	161,774	5,661	189,140
29	6,068	167,842	11,406	200,546
30	12,182	180,024	12,042	212,588
31	10,757	190,781	8,811	221,399
32	5,301	196,082	7,748	229,147
33	1,298	197,380	4,288	233,435
34	6,380	203,760	2,578	236,013
35	1,307	205,067	2,789	238,802
36	1,744	206,811	2,987	241,789
37	4,908	211,719	1,247	243,036
38	....	....	1,896	244,932

target. The job was progressing slowly, but Contractor E kept the gang down to the bare minimum despite continuous pressure to increase its size. This was a key factor in reducing potential lost time due to the slow pace of the structural work. The seventeenth, eighteenth, and nineteenth months were taken up by a general construction strike, which was finally settled, and work resumed.

Between the ninth and the sixteenth months, with a small work force and a large site, Contractor E was able to overcome the impact of the slow schedule by concentrating on operations that could be started and maintained at a reasonable work tempo. This changed after the nineteenth month, because the pressure was on to speed up the work and increase the size of the gang. It was now necessary to work all over the place in order not to delay the overall project. Pressure was exerted to increase the size of the gang to make up for previous schedule slippage and to speed up the work. Worker-hours consumed by the work began to overrun the worker-hours earned. The period between month 20 and month 30, which represented 25 percent of the total job duration, actually consumed 50 percent of the total worker-hours. It was during this period of acceleration that most of the worker-hours were lost. The project staff were aware of the loss. The site organization was very good, and great effort was put into planning, coordinating, and managing the work. Yet worker-hours continued to be lost. Job factors over which Contractor E had little or no control were having on impact on the work. These factors affected productivity, with a resulting loss in worker-hours. A record was kept of the following job factors that had an impact on the contract.

1. The structural work was badly behind schedule. The concrete was not poured according to schedule. Forms were not removed quickly enough, and areas were not made available to work in as scheduled. The blockwork and ceilings were neither started nor completed on time. When the topping was poured, the conduits were not protected from filling with water. Many floors were not poured level. Lines and levels given on plans often conflicted with those actually found on site.

2. Conduits installed under a previous contract were not identified, and many were found to be blocked. This affected the pulling of wire, which was part of the contract.

3. Controls, which were extensive on this project, were neither installed on time nor properly identified by the process contractor. Also, there were serious delays in the installation of equipment supplied by others. This interfered with the planned program to wire up and connect this equipment.

4. There were continual change orders, which added up to a 20 percent increase in the original contract amount. Considerable time was spent to track down missing information and to reconcile contradictions between plans and specifications. The 180 electrical drawings issued for this project averaged 3½ revisions per plan. Some had as many as 11 revisions. Only 20 percent of the plans had one revision or no revisions at all. Some revisions to drawings are to be expected in a contract, but when they become too extensive, they make proper job planning almost impossible, with a resulting impact on productivity. Approvals were late in coming, and this had the effect of further delaying the work.

5. There was inadequate coordination between the work of the sub-trades and the general contract work. Large spaces were left open and large areas were kept restricted for long periods of time, in order to accommodate future large equipment and large piping. This forced Contractor E to work out of his planned sequence and to start and stop many operations.

6. There was insufficient temporary heating during the winter months, with the obvious impact on the productivity of the workers. Much of the conduit installed on this project was polyvinyl chloride (PVC), and the lack of temporary heat affected the installation of this conduit because of the contraction and expansion of the material.

7. Priorities kept changing, and these changes had to be incorporated into the planned work schedule, with resulting disruptions.

Up to this point the analysis has concentrated on the loss of worker-hours due to job conditions and design changes. Contractor E suffered additional losses on this contract. If you adjust the 16-month overrun by allowing for time lost by the strike and added by the changes, the job was faced with a net overrun of nine months. During this period Contractor E was prevented from using his staff, equipment, and tools on other projects. Thus he could not earn the additional overhead required to cover the operation of his business. He was faced with additional financing costs for the extended time that payment of the holdback was delayed. Because of the extended duration of the project, some of the work had to be done at a higher wage rate than what was figured.

OVERRUN OF WORKER-HOURS

By far the largest portion of the losses suffered by Contractor E was the labor-cost overrun. Half of the labor-cost overrun was due to the reduction in productivity of the workers because of the impact factors. About 20 percent was for additional supervision, and 10 percent was for the increased hourly rates due to the extended duration of the contract. The balance of 20 percent covered adverse conditions, weather, and labor problems. Material costs were not substantially affected, but direct job expense, financing costs, and general overhead were all increased due to the extended duration of the contract. Contractor E kept detailed records and regularly informed the general contractor in writing about the factors that were having an impact on the work and increasing the costs. The replies that he received invariably maintained that these factors fell under job conditions.

The average construction manager or project engineer—and even the electrical contractors themselves—too often accept the loss of worker-hours and the other losses described above as a fact of life, as something covered under job conditions. The fact is that you have the right to run your job in accordance with your targets in the most efficient and productive manner possible. When you are prevented from doing so, you may be suffering damages. You should be aware of and understand what is happening, so that you can take the necessary steps to protect yourself. You should understand what constitutes a change to your contract and your rights to recovery under the contract.

WHAT IS A CONTRACT?

A contract is a legal relationship between at least two parties, creating obligations between one and the other which can be enforced by a court of law. The contract is essentially an agreement between the parties, in which the contractor offers to perform a given scope of work and the owner, or the owner's agent, or the general contractor accepts both that offer and the obligation to pay for the work done in accordance with the contractual agreement.

Usually, a building contract is made between an owner who calls for the work to be done and a contractor who carries out the work. It is common for the contractor to let out portions of the work, on a trade basis, to subcontractors. Subcontractors cannot look to the owner for payments, even though the work benefits the owner, because they have, in this case, no contractual relationship with the owner. They must look to the prime contractor for payments, and their rights are confined to making a claim against the prime contractor, except for the protection of their rights under the Mechanics Lien Act. The subcontractor is often bound by the general conditions of the prime contract.

The contractor is entitled to carry out the work without interference by the owner or the owner's agent. Neither owner nor agent can dictate the manner or order in which the work is to be carried out, unless the contract specifically calls for the directions of owner or agent to be followed. Contractors are responsible for scheduling the work and have the right to run the job as they see fit. They have the contractual right to control the work and to direct and supervise it on the basis of their skill and experience. They are responsible for the means and methods of doing the work, for the sequence of the work, and for the techniques and procedures which are used to control and coordinate the contract.

Architects and engineers who are agents of the owner cannot act beyond the authority given them by the terms of the contract. Their duty is to ensure, in a professional and impartial manner, that the terms of the contract are carried out. Cooperation and agreement are implicit in a contractual relationship, but the contractor must understand that the power of architects and engineers cannot be arbitrary or unreasonable. They cannot demand that the contractor act in a way that would involve additional cost not implicitly called for by the terms of the contract.

Particularly in a fixed-price contract, it is the duty and right of the contractor to do the work in the most productive and economical manner, as long as he or she complies with the intent of the design and with a reasonable work schedule. The experience of a contractor, and the contractor's knowledge of how to get the work installed, must be respected as long as the contractor's decisions do not interfere with or change the intent of the design on which the contract was based.

Agreements must be put into writing and must be unambiguous, precise, and clear. The intent and meaning of the contract must be read as a whole, and portions cannot be taken out of context. If there are disagreements as to the interpretation or meaning of clauses in the contract which cannot be resolved by discussion, then an arbitrator or a court may be the next resort to determine the intentions of the parties as expressed in their written agreement.

All contracts call for the following obligations:

1. The owner will make available the building site and all information, details, and approvals on time for the contractor to do the work efficiently.
2. The contractor will perform the work in a good and workmanlike manner, so that when completed, it will be suitable for the purpose for which it is required.
3. The work will be completed in a reasonable time with reasonable diligence, or else in accordance with the specified completion schedule when so called for in the contract.

Misrepresentation or wrong information may be issued innocently, or through negligence, or by intent. This can seriously damage your cost. Insist on precise information or data in writing. If in doubt, challenge what you consider an incorrect or unreasonable statement or demand.

An architect or engineer cannot refuse to certify work if it is properly completed. Failure to pay legitimate progress payments when due constitutes a breach of contract. Owners cannot refuse to pay for work substantially completed, because of minor defects, but you must rectify defects.

If you are prevented from performing your work within your contract by acts of the owner or the owner's agents, then you are released from contracted completion obligations and any associated penalties covered by the contract.

Too often the parties to a contract take a fatalistic attitude to the cost and time overruns resulting from breaches of the contract or damages inflicted by one party on the other. The magnitude of these impact costs is forcing contractors to be more conscious and cognizant of their rights under the contract, particularly their right to recover damages.

DAMAGES

The purpose of a claim for damages is to put the injured party, so far as money can do so, in the same position as if his rights had not been violated.* You must be able to prove that the loss was the result of a breach of your contract, and you must have the facts and figures to establish the magnitude of your loss. This is easier said than done. Most contractors do not keep adequate, comprehensive cost data and work records to specifically identify and quantify this type of loss. However, you are obligated to do all you can to limit and minimize the damages.

Since it is difficult to establish what damages an owner will suffer if a contract is not completed on time, a payment of a stipulated sum per day is often called for in the contract to be paid by the contractor to the owner for the delay. This arrangement is called *liquidated damages.* In order for the owner to be entitled to the liquidated damages, he or she must prove that the contractor was clearly responsible for the delay.

*"The right of the appellant would be to recover such amount of damages as would put him in as nearly as possible the same position as if no such wrong has been committed, that is, not as if there had been no contract, but as if he had been allowed to complete the contract without interruption" (Lord Cranworth, *Ranger v. Great Western Railway Co.*, House of Lords, England, 1854).

NOTICE PROVISIONS

You often fail to pay adequate attention to the notice provisions in your contract. The purpose of a written notice is to protect you, and to warn the other party of possible increases in cost so they can mitigate them if possible. This is a contractual and legal requirement that you must be very conscious of. Put it in writing. You must warn other parties that they are damaging you, and you must give them the chance to rectify or mitigate the problem.

CHANGES IN THE WORK

Usually the contract covers the manner of handling and pricing changes. It is implied in a contractual agreement that the changes will be of a minor nature and will not be such as to change the type of job for which you contracted.

A proportionally small amount of extra work which becomes necessary for the completion of a contract can usually be incorporated into the work without too much impact on the contract. However, extensive changes, or work which does not fall within the scope of the contract, will have a profound impact on the contract, and the contractor is entitled to recover any additional actual costs.

In establishing the value of a change or a claim there is an unavoidable conflict of interest between the contractor and the owner. The contractor is claiming for the actual cost of the work based on the particular job factors and conditions. The owner, on the other hand, is interested in paying only for the market value of the work received. Unfortunately, the market value of the change may be considerably less than the actual cost to produce this change based on the particular job circumstances. Nonetheless, a cost is reasonable if, in its nature or amount, it does not exceed that which would be incurred by an ordinary prudent person in the conduct of a competitive business. Cost must therefore be calculated on the actual and particular job circumstances, and not in relation to average or published cost criteria at large.

JOB FACTORS

When general cost units are utilized from estimating manuals or costing systems, care must be taken to apply the necessary factors to tailor the average units to suit the particular job conditions.

Although bidding time is usually very short, a contractor is held to have visited the site, to be aware of all local conditions and of the character of the work, and to assume all risks deriving from the work. In dealing with changes and claims, the question arises as to what the contractor should have reasonably foreseen in preparing the tender. This is discussed in greater detail in Chap. 7. It is obviously very important to use judgment and to clarify potential problem areas as much as possible before the contract is signed.

An act of God is a circumstance which human foresight cannot provide against, and it cannot be expected that a contractor should have foreseen it. Force majeure covers a wider class of events than an act of God, mostly events of human origin, such as strikes or wars, which are beyond the contractor's control. Strikes are usually treated as excusable delays, for which the contractor can get an extension of time but no additional compensation.

Contractor E, in the case described earlier in this chapter, suffered some losses for which he could not claim. Prior to the strike there were work slowdowns. There was the period of the strike itself. Also, there were problems inherent in the nature of this type of project, which he had foreseen and covered at the time of tender. He recognized that losses flowing from these factors were his responsibility.

However, there were factors over which he had no control and which were not his responsibility. They affected the work and lowered productivity because they forced the work to be done out of sequence and in a start-and-stop manner. They violated his contractual rights. These factors are summarized as follows:

1. Unavailability or late availability of areas that were required for scheduled work operations
2. Late receipt of drawings and instructions
3. Late receipt of approvals
4. Too many changes and interference of changes with the planned work program
5. Late delivery of equipment and systems supplied by others or by the owner
6. Interference by other contractors and inadequate project coordination
7. Insufficient temporary heating and insufficient workers' facilities

The contractor suffered damages from the above factors mainly in the form of lost worker-hours, but also in the form of additional job expenses and general overhead. The following is a summary of the damages incurred:

1. A decrease in productivity, so that it took more worker-hours to do a given amount of work

2. Increased labor rates for a portion of the work
3. Inefficient use of supervisory personnel and an increase in the cost of supervision due to the extended duration of the job
4. Added cost for estimating, engineering, material handling, and other direct job expenses
5. Increased expenses because tools, shacks, and construction equipment were tied up for a longer period than had been estimated
6. Additional expense due to extended warranty and bonding
7. Additional financing costs because of late payments due to late approvals and delay in the release of holdback
8. Added cost to cover the shortfall in overhead recovery due to the extended duration of the job
9. Higher material costs in some cases due to the changes and delays

If you are confronted with impact factors such as those described in this chapter, you must do what Contractor E did. He gave written notice as required by his contract, he kept records as detailed as possible to show the cause-and-effect relationship for the various damages, and he submitted a detailed claim for the damages which he had suffered.

7
Claims

OBLIGATIONS AND RIGHTS

When a hold is placed on all or portions of your contract, or if your contract is delayed by the impact of such factors as were detailed in Chap. 6, then your normal work schedule is disrupted. You are forced to start and stop many operations and to move frequently from one location to another in order to generate work for your crews. This results in a loss of productivity. It also results in an increase in the cost of supervision and in direct job expenses.

In practical terms, it is difficult to send away job forces which have been trained over a long period and are familiar with the work. Much time is often lost waiting for the hold to be removed, or for the delay factors to be eliminated, or for the tempo to pick up. Too late you realize that this is not happening and you must take steps to either relocate the crews or reduce their size. All this becomes clear in retrospect, when you analyze and try to explain why there was an overrun of the target worker-hours.

The electrical trade, like many others, has special problems. Not only are you involved in the roughing in and general progress portions of the job, but you are intimately involved in the finishing stages as well. When you work on fixtures, you are affected by the progress and problems of the ceiling trade. In order to install your switches, receptacles, and plates you are affected by the partition trades, plasterers, bricklayers, painters, and so on. Any disruption in the work of these trades and any architectural changes will affect you in terms of labor cost, even if there are no electrical changes involved.

EXTENDED DURATION

To do a complicated and comprehensive installation calls for an experienced and skilled staff. If you think about it, you will realize that in fact you have rented out your organization and staff to the particular project for a definite period of time, and you expect to be compensated by the project for the use of it. If your organization and staff are tied up on this job for longer than the contracted or normal period, because of actions by the owner or by others, then they must pay you additional rental charges, commonly called *extended duration charges.*

For example, if you rented an apartment and wanted to make use of it after the end of the lease, you would have to negotiate and pay the rental for the additional time involved. You might even be confronted with a penalty charge, since you have possibly prevented the apartment from being rented to a new tenant. By the same analogy, if you didn't make use of your apartment for a number of months, you would still have to pay full rental, since the apartment is there for your convenience any time you want to or have to make use of it.

This type of contractual obligation is true of most transactions. You don't expect to borrow money from a bank without paying interest. Even if you are an important customer of the bank, you don't expect to have the right to walk into their vault, help yourself to the money you need, and tell them, offhandedly, that you'll settle up at some time in the future. It doesn't work that way. You have to follow procedures, sign papers, and pay interest.

Yet in construction it happens all the time. Owners and their professionals very often feel that they can tie up your organization for extended periods or require you to do extra work, before having settled the conditions and terms of payment. There is a singular lack of sensitivity to the hardship and additional cost suffered by you when your organization and your money are tied up for longer than the contracted period, or longer than is fair and reasonable. This insensitivity is greatest with regard to impact costs which you suffer when your job is delayed or disrupted because of excessive changes or other factors and requirements imposed by owners and their professionals.

It is true that responsibility is a two-way street. When you are at fault, you have to pay for the consequences. But when the impacting job factors are out of your control, then you are entitled to be compensated by the parties who are at fault. Owners and their professionals must understand that when they introduce changes, particularly when the work is in progress, or when they change the conditions under which you must do your work, then they bear a responsibility for the resulting additional costs. It is often difficult to see and identify these increased job costs. Job costs are like a big pot of soup. All the ingredients having an impact get dissolved, and it is difficult to identify any given one. Yet it is obvious that labor costs, direct job expenses, and overhead keep mounting as long as the job is in progress.

There is a lot of resistance even to acknowledging a claim, and more still to paying for it. Since truth, like beauty, is in the eye of the beholder, the normal reaction is to disclaim responsibility, to put the blame on you, or to fall back on job conditions. In an industry as undisciplined as construction, the purchasers of your services expect a lot from you. They expect you to be able to foretell the future so that you will adequately cover yourself for all possible job conditions, disruptions, and changes and at the same time to come up with the low price bid to beat out all your hungry competitors. They also expect you to price changes in the same manner as your original bid, without taking into account the disruptive impact that changes have on the work in progress. You, in turn, try to pad the accounts, when you have the opportunity, so that you will cover your real costs.

It would be much better if both sides understood the real nature of cost and its inseparable relation to job conditions. Energy should be directed toward improving the design and improving the job conditions as logical steps toward decreasing costs and reducing claims. The real villain is not the contractor's price or claim, but the monumental waste that results when teamwork does not prevail to manage and control the factors that usually have an impact on the work. The emphasis should be on preventing the disease rather than arguing over the cure.

In the meantime, if you have a claim, you must establish the fact that you are entitled to it. Entitlement is based on proper and detailed records. If you can show the relationship between the cause and the subsequent effect and if you can quantify the effect, so much the better. The following job records are a must in this regard.

CLAIM FILE

At the beginning of every contract, set up a claim file. A claim file is like an insurance policy. You hope that you will not have to use it. However, if you suffer damages, it will give you ready access to the information required to back up your claim. All items which affect the progress of the job and which have a cost impact are stamped with the date and filed in the claim file. All memos, letters, field instructions, holds, and other records concerning acts of the owner, the owner's professionals, and the other trades that have an impact on your work—that delay or disrupt your work—are filed in the claim file. You thus have a history of the events that have an impact on your work.

DAILY LOG BOOK

The project manager maintains a daily diary in which to log all the pertinent events that take place on the job and the problems as he or she perceives them. This log book will give a picture, in capsule form, of how the job is progressing, what was worked on at any given time and where the work took place, what were the problems, constraints, and delays, and any other pertinent observations.

FOREMAN'S WEEKLY PROJECT REPORT

Each foreman fills out a project report on a weekly basis describing the operations on which that foreman's installers are working, the location where the work is taking place, and any problems which may have been encountered. This information is very useful for job control and progress billing, and as a record of delay, impact, and scheduling problems. A report form is shown in Fig. 7.1. It can be filled out neatly and briefly as follows:

1. Under "notes and comments," list the pertinent and applicable factors which have affected your work, especially factors beyond your control, such as site conditions, weather conditions, strikes, power failures, and lack of information or approvals.

2. Under "personnel on site," list the number of workers according to the various categories called for.

3. Under "description of the work," list the operations that your work force has worked on, along with the areas where the work took place, and mark in the percentage of the work completed for each operation up to the report date.

4. Under "areas or operations being delayed," list pertinent factors affecting your work, such as block walls not started or progressing too slowly; floor not poured; material piled up on the floor, preventing erection of scaffolding; vital material not delivered; change order affecting work not approved; area put on hold (indicate by whom); or lack of pertinent information (identify what it is). Be precise and identify the problem clearly, giving dates when possible. This is an important record. You may be blamed for delays that are not your fault. A record of what really happened, and when, will

---

Weekly Project Report

Job name _____   Job no. _____   Date _____

Notes and comments (use back of sheet if necessary)	Number of personnel on site
	1. Staff
	2. Foremen
	3. Electricians
	4. Helpers
	5.

Description of the work	Areas worked in	Percent completed

Delays

Areas or operations being delayed	Reasons for delays

Important deliveries	Project manager or foreman
	Name
	Signature

FIGURE 7.1 Weekly project report form.

very often settle the matter. Make this record straight and to the point.

5. Under "important deliveries," list the arrival dates of your major equipment and also those of other trades that affect your work.

These reports should be filled out weekly in a careful and considered manner, and a copy filed in the claim file. Information that is not recorded in an organized manner when it happens is often forgotten or becomes fuzzy or distorted. Write important information down and be sure that it will be available when you need it.

FORM FOR TOTAL OR PARTIAL STOPPAGE OF THE WORK

The form shown in Fig 7.2 is filled out every time a foreman is forced to increase or decrease the size of a gang, or move it from one location to another, for any of the following reasons:

Total ☐ **Stoppage of Work**
Partial ☐

Job name_____ Job no. _____

Date and time of stoppage _____

Location (be exact) _____

Description of the work _____

_____

_____

Reason for stoppage or movement _____

_____

_____

Loss of worker-hours suffered due to displacement of the crew, tools,

scaffolds, etc., to another location or task:

   Number of employees _____

   Hours lost (total) _____

Indicate other losses, if applicable _____

_____

_____

Date and time of resumption of work:

   Date_____ Time_____

Loss in worker-hours for new start-up _____

   Foreman _____ Project manager _____
           Signature   Signature

                                Date _____

FIGURE 7.2 Form for total or partial stoppage of work.

- A hold is put on the work by the general contractor or others.
- Work cannot proceed because of missing information or late approvals.
- Work cannot proceed because of interference or incomplete work by other trades.
- Work cannot proceed because of power failures or any other delaying factors.
- Work is stopped because of changes.

Each time that a worker or a crew is moved or stopped from working there is a loss of worker-hours. Each time this happens, the foreman fills out this form as a direct cause-and-effect record of time lost because of an impact factor. List precisely and accurately the type of stoppage or delay, where and when it occurred, and the number of workers involved. Multiply the number of workers by the worker-hours lost per worker to arrive at the total worker-hours lost for this stoppage. A copy of each stoppage-of-work report is sent to the general contractor or construction manager, and a copy is filed in your claim file.

JOB EXPENSES

Keep updated records of your job expenses. You may require this information if you decide to submit a claim for damages in the event that your contract is extended beyond the contracted or normal duration. This record is updated on a monthly basis, on a form similar to that shown in Fig. 7.3.

Keep a record of the tools used on the job and assign a fair market rental charge to each, so that you have a record of the value of the equivalent monthly rental for the tools used on your project. Updated copies of this and the other direct job costs are filed in the claim file. Aside from the obvious importance of having this information in the event that you must file a claim, it is also necessary for your organization to be aware of the monthly cost of running your job in terms of direct job expense. Expense is time-related, and time is money. This is an important lesson for your field people to learn.

Most claims have to do with delays and disruptions of the schedule. They are the most important impact factor. The delayed or disrupted work schedule is the most important cause. The most important effect is the loss of productivity. Loss of productivity means loss of worker-hours. The cost of the overrun of worker-hours represents about half of the amount that contractors attempt to recover in the average claim. Loss of worker-hours is the difference between the worker-hours actually consumed, as confirmed by payroll records, and those that you

Job Expense Report

Date _____

Cost per month

Job office staff (above field foreman)

Project manager   _____

General foreman   _____

Engineer/coordinator   _____

Estimator   _____

Storekeeper   _____

Job clerk/timekeeper   _____

_____   _____

_____   _____

Rental value job office and warehouse   _____

Office expense items

Telephone   _____

Telex   _____

Temporary power   _____

Office supplies   _____

_____   _____

_____   _____

Truck deliveries   _____

Clean-up expenses   _____

Bonding   _____

Rental value of tools   _____

FIGURE 7.3   Job expense record form.

estimated or targeted for doing the work and that you would have achieved had there been no interference.

The party against whom your claim is made will no doubt maintain that the loss of worker-hours is your own fault, that you didn't control your job adequately, that your workers were not producing, or that your original estimate was not accurate. Proving the loss of worker-hours is very difficult. You require supporting data.

PRODUCTIVITY CHECK

It is important to periodically check the productivity of your crews on the operations where most of the worker-hours are usually lost, such as conduit installation. Your target, based on your estimate and previous experience, may be to average 8 worker-hours per 100 ft of ¾-in rigid conduit on a given operation.* You find that your crew is averaging 13 worker-hours for this work. You should check and analyze the reason for this drop in productivity. If you conclude that job factors are interfering with your work, then you should notify the general contractor or construction manager of this fact, and go on record as intending to claim for the lost worker-hours. Keep a record of your productivity checks and document them properly as to areas, impact factors, and dates. Such as-built information is a lot to ask for, but there is nothing like field data to help bolster your claim, particularly if you can establish a relationship between cause and effect and can quantify the resulting loss.

CASE HISTORY OF A CLAIM

The need to keep good records is illustrated in the following case. Electrical Contractor E obtained a contract in the amount of $4.5 million for the electrical work in a large automated postal facility designed to handle parcel and letter sorting for a large metropolitan area. This was a "fast-track" project, and the contract which is reviewed here was one of many packages awarded to different contractors. The scope of work included lighting, secondary distribution, electric heating, and motor connections. Process equipment and other electrical packages were handled under separate contracts. The work was scheduled to be completed in 25 months. Some areas were singled out as priority areas, with milestone dates listed in the specifications. As an example, the office area was scheduled to be completed in 13 months. In fact, even though the overall completion date was met, the individual milestone dates were greatly delayed. In the case of the office area it was not completed until month 25, which represented an extended duration of 12 months for that part of the contract. The individual milestone dates were not met, and portions of the work had to be accelerated in order to meet the overall completion date.

ACCELERATION

Most projects have a construction period which is either normal for that type of installation, or is written into the contract documents. Some-

*U.S. Customary System (inch-pound) units of measure are used throughout this book. Please refer to the appendix, where a table of conversion factors is provided for readers using SI (metric) units.

times this schedule is accelerated in order to make up lost time or to accommodate revised occupancy requirements of the owner.

When this occurs, you will be confronted with additional costs and expenses. These costs include, but are not limited to, the following:

- A drop in efficiency and productivity
- Additional supervision
- Additional coordination
- Additional procurement expenses
- Impact costs due to overstaffing and overtime
- Possible unfavorable effects on other jobs and on your estimating program, as you are forced to move workers, supervisors, and head-office personnel to generate the accelerated program

An accelerated program forces you to increase the size of the work crew beyond economical proportions. You may have to bring in additional foremen, who are thus withdrawn from other company duties. Under a normal, planned work schedule you attempt to select the best personnel for optimum performance. In an accelerated program you are often forced to hire workers from the labor pool, many of whom may be inexperienced or below your standards.

An accelerated program may decrease the efficiency of your work because many subtrades have to install their work at the same time, which can play havoc with your coordination. You may be forced to install your work before you are ready, and often in smaller quantities at each time, in order to allow the other trades to proceed with their work. Instead of starting and completing an operation, you may be forced to come back many times before it is completed. This results in the loss of time.

Most serious of all, you may be forced into overtime. *Overtime should be avoided when at all possible.* It is generally counterproductive, as explained in Chap. 9, and has serious cost effects on your contract if it persists for any length of time. If you are forced into an accelerated program, keep track of the additional expenses resulting from

- Loss of productivity, as described above.
- Extra worker moves due to start-and-stop work operations, as you are forced to deviate from your coordinated work program.
- Extra supervision.
- Extra coordination and expedition of work to obtain information and materials ahead of your original schedules.

- Extra trucking and material handling to keep up with the rush nature of the work.
- Extra cost of tools.
- Extra cost of storage.
- Extra cost due to overtime and its impact on the balance of the work. The premium portion is the tip of the iceberg that is visible. You also have to account for the lost productivity, as illustrated in Table 9.1.

NOTICE PROVISIONS

In all claim situations, you have to inform the party who is damaging you in writing, by memo, letter, or registered letter. Describe what is happening, and give the other party the chance to improve the situation. Copies of these letters and memos should be filed in the claim file, so that you have the complete picture and all the supporting data in one place, where they are readily available. Contractor E sent out 180 letters and memos regarding items that were interfering with his contract. These letters and memos fell into the six general categories shown in Table 7.1.

OVERRUN OF WORKER-HOURS

Contractor E was confronted with acceleration as well as with other job factors that increased his cost. Table 7.2 is a summary of worker-hours actually spent on his contract, as compared to the worker-hours estimated.

You can see that Contractor E suffered an overrun of 74,000 worker-

TABLE 7.1 CASE HISTORY OF A CLAIM: SUBJECTS OF LETTERS AND MEMOS

SUBJECT OF WRITTEN COMMUNICATION	NUMBER OF WRITTEN COMMUNICATIONS
Clarification of design problems and technical questions	40
Coordination with other trades, obstructions, and interference	35
Delays in issuing drawings, revised drawings, and approvals of drawings	30
Impact of changes and late approval of changes	30
Delays to work because of schedule slippage and late completion of critical structural and building elements	30
Adverse site conditions, approvals of materials, etc.	15
Total	180

TABLE 7.2   CASE HISTORY OF A CLAIM: ACTUAL AND ESTIMATED WORKER-HOURS COMPARED

ITEM	DIRECT WORKER-HOURS	INDIRECT WORKER-HOURS	TOTAL WORKER-HOURS
Total worker-hours consumed	214,000	63,000	277,000
Less adjustment for extras	37,000	12,000	49,000
Worker-hours for basic contract	177,000	51,000	229,000
Estimated worker-hours	127,000	27,000	154,000
Overrun	50,000	24,000	74,000

hours on this contract, which works out to about 26 percent. Table 7.3 is a tally of the worker-hours actually spent as compared to the worker-hours estimated for the key operations, namely, conduit installation, wiring, cable trays, Teck cables,* and fixture installation.

The major loss of worker-hours occurred in the conduit work. This operation was the most affected by changes, lack of coordination, waiting for answers to technical questions, site obstructions, and acceleration of various portions of the work. However, installation of cable trays and Teck cables suffered proportionally the greater worker-hour loss because of the critical impact that the various conveyor installation contracts had on this operation.

Contractor E was well-qualified for the work called for in his contract, with a history of excellent, on-time performance. This project, however, was beyond his control because of the numerous changes and the lack of adequate interfacing and coordination between the numer-

TABLE 7.3   CASE HISTORY OF A CLAIM: ACTUAL AND ESTIMATED WORKER-HOURS COMPARED FOR KEY OPERATIONS

OPERATION	TOTAL WORKER-HOURS SPENT	WORKER-HOURS SPENT ON EXTRAS	CONTRACT NET WORKER-HOURS	ESTIMATED WORKER-HOURS	WORKER-HOURS OVERRUN
Conduit	91,000	14,000	77,000	50,000	27,000
Wiring	16,000	4,000	12,000	10,000	2,000
Cable trays	23,000	3,000	20,000	10,000	10,000
Teck cables	22,000	5,000	17,000	7,000	10,000
Fixtures	21,000	2,000	19,000	18,000	1,000
Subtotal	173,000	28,000	145,000	95,000	50,000

*A Teck cable is an armored cable with a PVC jacket overall. The armor can be steel or aluminum.

ous contract packages, many of which were tendered too late. This made proper planning and coordination very difficult. The history of fast-tracking is littered with jobs that suffered extensively because of late delivery of drawings and information and because of poor interfacing and coordination of the various contracts. The construction managers often try to protect themselves with clauses stating that coordination is the responsibility of the contractors. All contractors are responsible for coordinating their own work, but it is the responsibility of the construction manager to make up a master schedule and to manage the project in accordance with a master plan. In this case, the master plan was not revised or updated to take into account the changes that were being made and the new contract packages that were being awarded. Contractors on site were not aware of what further packages were to be tendered or when they were to be tendered. Contractors tendering subsequent packages based their bids on data and schedules which the construction manager should have informed them were already obsolete.

This state of disorganization led to installation requirements—particularly those pertaining to the location of runs and fastening methods—that were so abnormal, so unreasonable, and often so bereft of good common sense, that it had a very damaging and demoralizing effect on the installers. Contractor E urgently required hard information about the installation of the conveyors and their location, height, and size in order to proceed with his work. However, contracts for the conveyors were being awarded as design-build packages, and the pertinent information required by Contractor E was not available. In fact, some of the conveyor packages had not even been awarded when the tray installation was scheduled to proceed. As a result, Contractor E was forced to run the trays at excessive heights (over 50 ft) in order to clear the future conveyors under any and every circumstance.

The alternative was to be responsible for any subsequent interference with the location and functioning of the conveyors. This was too great a risk to take. The lack of control by the construction manager over the conveyor installers, particularly since these conveyors crisscrossed the entire project, made it almost impossible to plan and coordinate the work of many operations in an effective way. This resulted in a tremendous loss of worker-hours. On many occasions the work was accelerated in given areas because of the imminent arrival of conveyors. The acceleration often required the overstaffing of these affected operations. Then the conveyors would not arrive until months later. Large expenditure of worker-hours was required to set up the scaffolds for the tray installation, to move them, and to reduce their heights to clear the girders. Large amounts of additional set-up time were required to arrange the pulleys to pull in the Teck cables at such great heights. During these operations the floor was obstructed with equipment and parts of conveyors.

Contractor E had based his tender on running the trays at reasonable elevations, since the equipment which was fed by the Teck cables was at ground level. This was the logical way to do the work, and the way in which it would have been done had the conveyor information been available on time. The worker-hours lost on the conduit and tray operations were the direct result of poor design, too many changes, and poor or nonexistent coordination.

The wiring and fixture operations were not subject to the same impact factors. They were carried out when a good part of the interference had already been overcome. As a result, the worker-hour targets for these operations were almost met, as can be seen in Table 7.3.

If you assume that the average loss of productivity for the key operations shown in Table 7.3 applies to all operations, then the total loss of worker-hours (WHs) would be calculated as follows:

$$\frac{229,000 \text{ (contract WHs)}}{145,000 \text{ (key-operation WHs)}}$$

$$\times \; 50,000 \text{ (WH overrun on key operations)} = 79,000 \text{ WHs}$$

which works out very close to the actual loss of 74,000 worker-hours reported in Table 7.2.

When Contractor E commenced the work, the building schedule was already falling behind. Construction starts, of walls and ceilings that were scheduled to be started on given dates, were late. The problems stemming from the manner of awarding and dealing with the process equipment and conveyors have already been described in some detail. Construction data for these critical contracts were just not available on time to help Contractor E with coordination and installation. The sketchy and deficient information relating to the process equipment and conveyors resulted in continual coordination and interference problems.

During the course of the contract, 203 change orders were issued, which added up to 38 percent of the original contract amount. This created a stop-and-go work pattern that required the frequent relocation of workers. In order to quantify this loss as much as possible, Contractor E recorded the number of worker moves experienced in the operations listed in Table 7.4. If you assume a loss of 3 worker-hours for every move, then you can see that about 18,000 worker-hours were lost on excessive movement of workers in and out of operations. This is graphically illustrated in Fig. 7.4.

Figure 7.5 illustrates the effects on the productivity of various crews installing ¾-in rigid conduit, under the supervision of different fore-

TABLE 7.4   CASE HISTORY OF A CLAIM: RECORD OF WORKER MOVES

WORK OPERATION	NUMBER OF WORKER MOVES
Conduit up to 1¼ in	1849
Conduit 1½ in and up	506
Wire no. 14 to no. 8	878
Wire no. 6 and up	458
Teck cable	693
Cable tray	454
Wiring channel	658
Fixtures	496
Total	5992

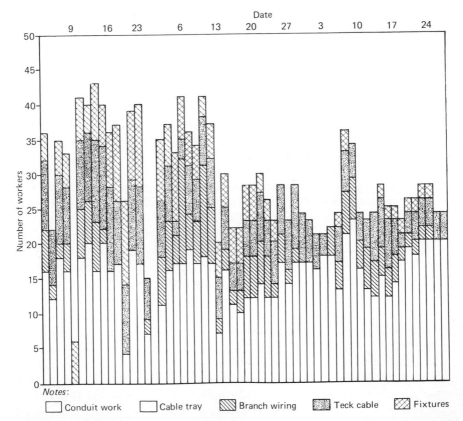

FIGURE 7.4   Tabulation of worker moves created by stop-and-go pattern of work.

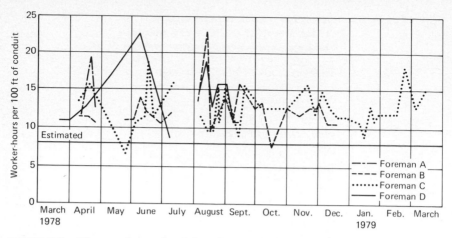

FIGURE 7.5 Changes in productivity of various crews caused by job conditions.

men because of delays, changes, and poor coordination. The estimate was based on an average productivity of 8 worker-hours per 100 ft. The actual productivity fluctuated enormously, around a mean installation rate of 13 worker-hours per 100 ft. This type of study was made for the other key operations and showed a similar drop in productivity.

In analyzing this case history, you can see that worker-hours will be lost, because of the impact factors previously described, even if you are prudent and do your best to control your job. In the case of Contractor E, approximately 20 percent of the worker-hours were lost because of the inordinate movement of workers into and out of operations. Approximately 30 percent of the lost worker-hours were consumed by additional supervision and administration. The balance of 50 percent was lost in the decrease of productivity because the tempo and rhythm of the work were constantly broken by delays, changes, and interference problems.

Lost worker-hours are difficult to identify and more difficult still to quantify with regard to cause and effect. It was therefore the purpose of this study to identify some of the areas where worker-hours are lost. Try to minimize excessive movement of workers into and out of operations. Keep records of the resulting work stoppages, additions to and reductions of crews, and movements of workers from one uncompleted task to another. Carry out regular productivity checks of the key operations, the ones on which worker-hours are being lost. By comparing the actual performance with what you estimated, by establishing the causes that are affecting productivity adversely, and by keeping proper job records, you will be in a better position to support a claim if it becomes necessary to make one.

8
Work Force and Worker-Hours

Those of you who follow sports know very well how preoccupied sports management is with personnel. There are farm teams, scouts, and a continuous search for better players and for stars. The sports pages of your newspapers are filled with the evaluations and ratings of players. No team can hope to succeed and win games without a proper personnel policy. What is more important, the players themselves fully understand the importance of performance and the need to continually improve in order to succeed.

The work force is the principal resource of your industry and your business. It is essential, therefore, to find the ways to develop this resource to its maximum potential. You cannot relinquish your responsibilities and rights to develop and control the work-force resource without putting your industry and your business in jeopardy.

PERSONNEL POLICY

Every contractor, large or small, requires a personnel department and a personnel policy. This department does not have to be very formal or very elaborate, but it should be run by a person with a capability and an interest in the field of labor relations and worker development. Programs to upgrade the skills and expertise of personnel managers in accordance with professional standards are vital. Doing what comes naturally is no longer enough in dealing with the work force, as witnessed by the steady decrease in productivity and the increase of labor-relations problems.

The tendency in the construction industry has been for worker performance standards to slide downward to ever lower common denominators. Important reasons for this are an erosion of motivation and a lack of adequate communication between management and workers. The composition and attitudes of the work force have been steadily changing. The average worker is now younger, less worried about job security, and better educated than the average worker of former years. Today's average worker is more concerned with job satisfaction. He or she wants to participate in some of the planning and decision making on the job, wants to feel part of a team, and wants to be recognized for good performance. Workers today want to be somebody. You have to adapt yourself to these new attitudes so that you can channel them into productive work performance.

In many areas, contractors cannot pick or choose their workers. They must very often take them from a common labor pool or from the top of the list at the union hiring hall. Union hiring halls give the union great power, which they can use to the detriment of a contractor. Particularly in the case of large jobs, the first workers from the union hiring hall may be hand-picked by the union. These workers, along with the shop steward, very often set the tone and tempo of the work. They are there for the duration of the job because of the principle of the first worker hired being the last one to be laid off. Peer pressure very often prevents workers, who for reasons of inner motivation might want to work smarter or faster, from doing so.

Under such circumstances it is not surprising that contractors and their foremen become frustrated. Discouragement and frustration unfortunately will push people to look for scapegoats rather than solutions. Contractors point to union stewards who refuse to put in a fair day's work and spend the day talking to the workers, one and two at a time, thus preventing them from working. This type of nagging problem can be overcome by better communication between labor and management. If you expect the other side to be intransigent and if you expect no good to come from communicating with them, then your expectations will no doubt be fulfilled. Expectations tend to be self-fulfilling. By developing labor-relations managers with more expertise and a more professional understanding of the role that management must play to solve labor problems, your expectations will become better focused and more readily attainable.

You have to start with the assumption that your average workers are interested in doing a good day's work—that they are responsible citizens who feel good when they know that they are being productive. If this is not the case, it is up to you to instill such an attitude in your workers. Communicate the necessary information for them to do their work properly. Set realistic worker-hour targets so that they will have a means of measuring their performance. Make it clear that worker-hours are monitored not for the purpose of judging or punishing them, but for the purpose of eliminating unnecessary lost time and increasing the efficiency of the work. In aiming to increase the average performance you must accept the fact that some people work faster or better than others, but on the average the performance must meet your targets. Encourage workers' ideas and suggestions of ways to improve performance and cut out wasted or unnecessary activities. Above all, don't hold them responsible for poor productivity if the real culprit is poor planning, poor tools, late material deliveries, or poor job management.

DEVELOP FOREMEN

The cornerstone of your personnel policy is to find and develop high-caliber foremen. Along with technical expertise, they require management expertise, which calls for appropriate educational programs, both in-house and industrywide. Don't take your foremen for granted, and don't leave it up to someone else to train them. Foremen usually come up from the ranks. They are workers who have shown that they can think and plan ahead. They are usually authoritative, have leadership abilities, and are able to communicate their ideas. When these people are singled out, they should be trained to acquire a management point of view.

PROCEDURES

A personnel policy involves specific hiring procedures and records for rating your employees.

The project manager or foreman directs a request for personnel to the personnel manager at the head office. This is the person designated by your company to be in charge of personnel. If he or she happens to be out of town or absent, the request goes to the paymaster, who will obtain the personnel manager's approval upon his or her return. Make the request in writing whenever possible. In this way you will be forced to think and plan ahead. Make the request at least one week ahead of your requirements. This will enable the personnel manager to review the situation on other jobs where competent workers may be available for transfer to your project. It will also give the personnel manager time to try to obtain the best possible workers to suit your needs.

When requesting personnel, specify the type of work that they are required for. Your request should establish whether you need workers for high-voltage operations, control wiring, heavy conduit, installation of fixtures, or finishing work. By specifying the type of work, there is a good chance that the personnel manager will be able to find you workers with that background and experience.

If you know of an employee on another project who would be better suited for a particular function on your job, take it up with the personnel manager, who will look into it. If you know of a good worker looking for employment, bring this up as well. Take particular care to search out good apprentices. The apprentices are your farm team, from which you will draw your future journeymen. Make sure that they are interested in the trade and that they acquire the proper experience and working habits.

All new personnel must report to the paymaster with the documents required for them to be hired in accordance with local regulations. They must be signed in by the paymaster before they report to the job. When an employee is transferred to another project, he or she is given a transfer slip and a copy is sent to the paymaster. Upon release from a job, the employee is given a release slip in a sealed envelope to be presented to the paymaster. Make sure that the slip includes the number of hours worked during the week. These hours should be phoned in to the paymaster in advance, to allow time for making up the pay check. The badge and hard hat which were issued to the worker when hired must be returned when he or she is released. At least one week's notice, and more when possible, should be given to the personnel manager before layoffs. At this time, inform the personnel manager as well if some good workers or any with exceptional abilities are included in the layoff.

Personnel Evaluation

Before a layoff or a transfer, arrange for the foreman to fill out a rating form for each worker involved. Mail these to the personnel manager at the head office for filing and future reference. The purpose of the rating form is to give an in-depth picture of the worker as perceived by the foreman. After workers have been rated by different foremen, you should be able to get to know them and learn about their characters, their attitudes, and their abilities to learn and develop in the trade. An evaluation of a worker should answer the following questions:

1. Does he possess average, below-average, or above-average intelligence?
2. Does she make an effort to improve her knowledge of the trade?
3. Is he a willing worker?
4. Does she show ambition to get ahead?

5. Does he have leadership abilities?

6. Is she taking extra courses or trade training?

7. Does he possess any special skills, such as welding, rigging, or mechanical?

8. Does she have expert knowledge of control wiring, high voltage, or electronics?

9. Does he have special proficiency in operations such as conduit work, fixture installation, or finishing work?

10. Can she read a blueprint?

11. Does he get along with his fellow workers and with management?

12. Does she enjoy her work?

13. Is he honest?

The evaluation should also list the work record, other companies for which the employee has worked, positions occupied, and the corresponding dates. Obtain a small, inexpensive picture of the worker and staple it to the form. You must be on a constant look-out for employees who show above-average intelligence, skill, and leadership abilities. Such personnel should receive your special attention and guidance to help them develop their talents and grow in your firm.

The personnel policy must take steps to elevate the level of competency and improve the attitude of all the employees. You will achieve this by the careful selection and training of your foremen and project managers. You have the expertise. What you need is leadership—informed leadership. Programs such as regular newsletters or informative meetings dealing with the proper use of tools and fastening methods will help, and they can be organized either in house or through your local contractors' association. Devise ways and means of communicating with your employees. The keys to achieving effective communication are good planning and good management. A well-run job, where operations take place as planned, where material and information are available on time, and where the installers know what is expected of them, will result in good morale and a desire to perform. At work, as with everything else, people react positively to a positive environment.

Hiring

1. Each project manager, or job foreman in the case of small jobs, has a supply of preprinted worker requisition forms, to be forwarded to the head-office personnel manager when additional workers are

needed. In the case of emergency requirements, the project manager or foreman can consult the personnel manager verbally by phone, but he or she is still required to follow up with the written request.

2. When the new workers present themselves at the jobsite, they must hand in their engagement slips duly signed by the personnel manager.

3. The personnel manager checks frequently with the paymaster that the above procedure is being followed, that the proper rates are being paid, and that there are no dummy employees on the payroll.

4. In the case of out-of-town jobs, the employees are hired on the jobsite by the project manager, who works closely with the head-office personnel manager and obtains his or her approval every step of the way. Copies of all documentation, engagement slips, and competency cards are forwarded to the head-office personnel manager for every employee hired, together with an inexpensive photograph which can be used for identification purposes.

Timekeeping

1. Larger projects have punch clocks and job clerks whose function includes timekeeping. The timekeeper on a large project prepares punch cards listing each employee's name and number. Punch cards are maintained for all employees, including foremen. The time clock should be placed in such a position that the timekeeper can survey the employees as they punch in and punch out.

2. Time cards are collected by the timekeeper and initialed by the foreman and the project manager. In order to prevent the punching in of late or missing employees or the insertion of dummy employees, the punch-in–punch-out process is scrutinized by the timekeeper and periodically by the project manager. The personnel manager visits the site occasionally to check that this procedure is being followed.

3. On small projects, the foreman usually prepares the time reports and forwards them to the paymaster. In order to prevent overreporting of hours or insertion of dummy employees, the personnel manager spot-checks the small jobs periodically, and also initials all the time slips.

4. In the case of out-of-town jobs, the personnel manager visits the site periodically to verify the procedure. All out-of-town time sheets are initialed by both the project manager and the personnel manager.

Termination of Employment

1. In accordance with most collective labor agreements, employees who are laid off must receive their final pay within 48 hours of their release.

2. The foreman or project manager signs a preprinted release form, which is sent to the personnel manager, who initials it and hands it over to the paymaster. This written notice is required to prevent released employees from remaining on the payroll. Final checks are issued only after the release form has been properly completed and signed. Whenever possible, the employees who have been laid off must present themselves to the paymaster to obtain their final checks. The employee's signature on the release form should be compared with his or her signature on file.

3. In the case of out-of-town jobs, a preprinted release form is signed by both the foreman and the employee and is immediately forwarded to the head office. Employees who are laid off are usually paid directly by the project manager from a fund provided for this purpose. Reimbursement for this payment is processed only after the release form has been reviewed and approved by the personnel manager at the head office.

4. Employees rehired within the same calendar year need only complete a preprinted reinstatement form and sign it. This form must be signed by the foreman and initialed by the personnel manager.

Accounting

1. The paymaster receives the time records and checks the hours. The payroll data are given to the machine operator, who prepares the checks and payroll journal simultaneously.

2. Checks are signed only after being processed completely. The signing officers should periodically verify the checks to ensure that the computations are correct. They should also verify that the gross pay as calculated by the payroll department agrees with the gross pay as calculated by the costing department. Signed payroll checks should not be accessible to any member of the payroll department under any circumstances. In order to prevent anyone in the payroll department from retaining checks made out in the names of dummy employees, the grouping for distribution of the checks should be carried out by an employee who has no connection with the processing of the payroll prior to signature.

3. Periodic surprise distributions of the payroll should be carried out by the personnel director. Unclaimed checks should be returned to the head office, and the affected employees must claim them there after presenting sufficient identification. The personnel director must investigate when checks are unclaimed for a period in excess of 2 weeks.

4. Time records are prepared in two copies: the original, which is forwarded to the paymaster, and a copy, which is forwarded to the costing department.

5. In the case of traveling allowances or errors in computation of hours or rates, the job sends the head office a preprinted payroll voucher form which contains the name of the employee, the date, the reason for the voucher, and the amount, including receipts if applicable. This form is signed by the employee and initialed by the foreman and the personnel director.

6. The bank reconciliation of the payroll is prepared by the accounting department and not by the payroll department.

BAR CHART AND MANPOWER GRAPH

You must find the time, at the start of a job, to plan and schedule the operations and to work out your personnel requirements. This can be easily done, graphically, in the form of an operations bar chart, from which you can derive a work-force graph, as shown in Fig. 8.1.

Start by calling a coordination meeting of the job estimator, the contract manager, the purchasing agent, the general superintendent, and the project manager. The estimator should then review the estimate and bring forth all the details and tendering information affecting it, which should be clearly understood by the installers and the purchaser. In this way the study, notations, and details which the estimator accumulated during the estimate will not be lost, but will be incorporated into the program to plan, schedule, and control the job.

The coordinator and project manager must review the plans and specifications thoroughly and list all the pertinent data that will affect the purchasing and installation functions. These must be noted down and inserted in the job reference binder or file for easy and continuous reference. While studying the estimate, the coordinator and job manager must then break down the contract into logical events or operations. These events or operations must be sorted out as to the probable sequence in which they will occur and their probable time frame for installation. The duration of every operation and the number of work-

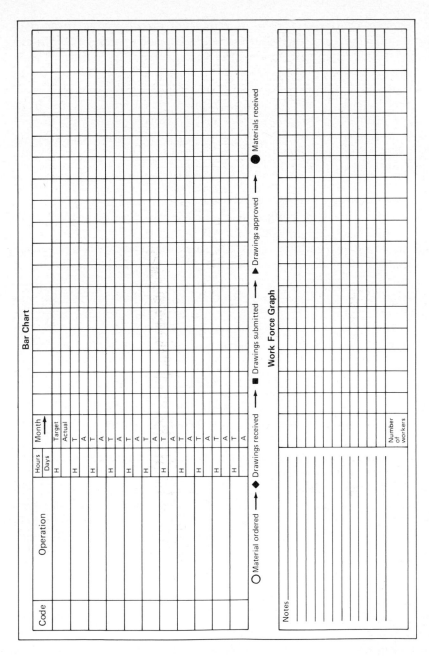

FIGURE 8.1 Bar chart form.

ers required are derived from an analysis of the estimate. These data must be sifted through the judgment and experience of the coordinator and the project manager, so that the various operations are not over- or underbudgeted on the basis of job conditions.

Estimating of necessity deals with labor units for average job conditions. You hope that the resulting total number of worker-hours from the estimate will be sufficient to do the job. However, in apportioning the worker-hours to the various operations, the average must be divided up, so that the difficult areas get the additional worker-hours and the easy, repetitive areas are cut down to suit.

Constraining activities, such as shop-drawing approval, placing of orders and delivery of materials, supply or rental of special tools and equipment, pouring of critical slabs, or construction of critical walls must be identified, scheduled, and marked on the chart. The general contractor or construction manager must be asked to supply a proposed construction schedule so that all of the above operations can be interfaced with the building program.

The bar chart can then be prepared along with the work-force graph which is derived from it. A ceiling should be established and drawn on the work-force graph to limit the maximum size of the labor contingent. The project manager must be instructed not to exceed this maximum work-force loading without permission from the head office. This is an important control, since job pressures very often result in larger gangs than the actual work budget allows.

The bar chart will have to be revised and updated periodically to reflect the developing job conditions. You must insist that your staff mark in the as-built data as the work progresses, so that you have a continuous and up-to-date comparison of the as-built performance with your planned performance.

The aim of scheduling is to organize and stack the operations in such a way that you can do the work in a continuous manner, without peaks and valleys and with the smallest gang possible. Every time your workers are shifted in and out of operations or areas, or your gangs are increased or decreased, worker-hours are lost, as illustrated in Fig. 8.2. It is the purpose of scheduling and job planning not only to identify the start and finish points of an operation and its magnitude in terms of personnel required, but also to study the relationship and interfacing of the various operations. Think of the work as being done in stages and the operations as being staffed by crews of optimum size. A given crew should be programmed to complete an operation in one stage and move efficiently on to the next. Crews can be split when smaller parallel operations are involved, and joined together or augmented for larger operations. With this approach you will be able to organize your opera-

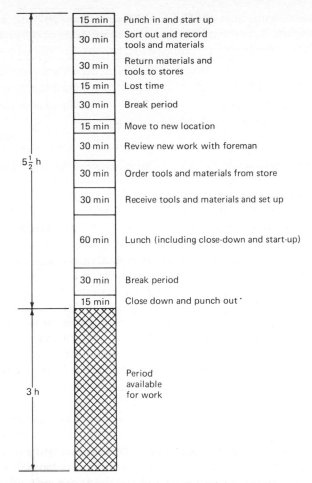

15 min	Punch in and start up
30 min	Sort out and record tools and materials
30 min	Return materials and tools to stores
15 min	Lost time
30 min	Break period
15 min	Move to new location
30 min	Review new work with foreman
30 min	Order tools and materials from store
30 min	Receive tools and materials and set up
60 min	Lunch (including close-down and start-up)
30 min	Break period
15 min	Close down and punch out ·

Period available for work

FIGURE 8.2   Time lost due to relocation from one job to another.

tions to maintain a steady flow of work, and cut down the movement and jumping around of your crews to a minimum.

The personnel manager should obtain copies of the work-force graphs of all the active jobs and superimpose them on a master graph. This will create a picture of the anticipated total work force which may be required to service all the jobs at any given time. By comparing the planned personnel requirements with the actual as-built personnel being utilized, the manager will be in a position to forecast future requirements and also to move employees from job to job in a logical and efficient manner. The principle here, as in every other aspect of the work, is to be aware of potential problems before they become actual, so that effective action can be taken in time. Since every project has critical and noncritical operations, the latter can be shifted about in the overall schedule to allow you to utilize your best workers and supervisors in the most effective manner possible.

GETTING THE MOST FROM WORKER-HOURS

Low productivity can result just as often from poor management on your part as it does from a poor attitude or poor motivation on the part of your employees. Good productivity very rarely develops spontaneously, without planning and effort; it must be strived for and achieved.

All personnel in your organization must develop the gut understanding that it is unacceptable to waste worker-hours, just as it is unacceptable to waste food or energy. The high standard of living which all desire and deserve cannot survive with such waste.

Make it clear to all concerned that every operation carries with it an optimum number of worker-hours based on the particular job conditions. These are the targets which you must set and strive to achieve. Their achievement requires that the operation be well planned and coordinated, that problems be worked out in advance, that the best fastening methods be selected, that material and information be available when needed, and that the crew be carefully chosen from the best workers available and in the correct number required.

The foremen must keep daily records of the amount of material installed each day by their teams or crews, and the number of workers and hours used in its installation. This is particularly necessary for such repetitive operations as conduit and tray installation, wire pulling, and fixture installation. The resulting productivity records will be a check on your estimations of labor units and on the work patterns and work habits of your various workers and foremen. The purpose is to achieve maximum efficiency and productivity together with a healthy, relaxed, and acceptable tempo of work. The form shown in Fig. 8.3 is used for this purpose. Alongside the name of the worker or the names of a team, if they are working as such, mark in the amount of material and the number of hours consumed on a daily basis. This study applies to critical operations such as conduit work and fixtures. If the work is done under abnormal or difficult conditions, add the code D to identify this fact.

LABOR UNIT EVALUATION

Just as regular exercise is indispensable in order to maintain a healthy heart and body, so is the need for regular productivity records to

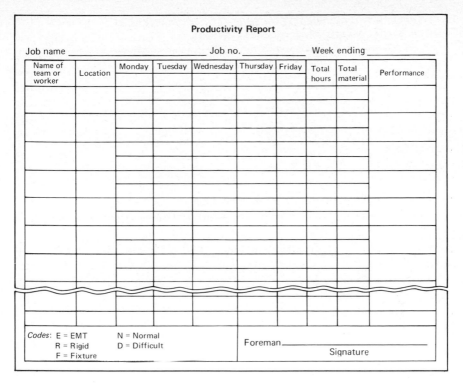

FIGURE 8.3 Productivity report form.

maintain top performance. The combination of the complexity of the average job and human nature will inevitably result in lost time if the work is allowed to develop spontaneously, without planning and without checkups. A program of continuous productivity evaluation of your critical operations, the ones that can make or break your job, will give you the necessary control to bring in your labor cost within your target.

Chapter 2 refers to field studies that show how the worker-hours are utilized during a working day. On the average, only 35 percent of the working day is utilized for actual installation activity. Therefore, getting the most out of worker-hours requires a style of work in which the productive 35 percent of the day is increased to 55 percent or higher. Since only a fraction of the day is used for actual productive work, it is doubly important to get the most out of every productive hour. The worker-hours, on an average project, are consumed as shown in Table 8.1.

You can easily see that the installation of conduit is the critical opera-

TABLE 8.1 **PRODUCTIVE ACTIVITIES AS PERCENTAGES OF TOTAL PRODUCTIVE WORKER-HOURS**

ACTIVITY	PERCENTAGE OF TOTAL PRODUCTIVE WORKER-HOURS
Conduit work	45
Wiring	15
Fixtures	15
Supervision	10
Auxiliary systems, motor starters, and finishing work	5
Balance of work, including temporary wiring	10

tion which can make or break your labor targets. Educate your foremen and installers so that when they look at, for example, 20 lengths (200 ft) of ¾-in rigid conduit, they will see work for a team of two for a regular eight-hour day.

9
Productivity

OPTIMUM WORKER-HOURS

Every operation carries with it an optimum number of worker-hours that are required for the work to be done under the governing job conditions. This quantity of worker-hours, as far as you are concerned, is based on the number of worker-hours that you estimated or targeted for this operation. The ratio of the optimum number of worker-hours compared to the quantity actually consumed is the percentage, or factor, by which you judge the productivity. If you estimated 40 worker-hours to complete an operation, but it actually took 80 worker-hours, then you would say that the productivity was 50 percent of what you figured.

Many industry sources have made extensive studies and have come up with average labor units for various types of job conditions. These data, together with the feedback which you should get from your jobs on a regular basis, will give you the means for formulating labor units that represent the optimum productivity you should be striving for. Without these optimum labor targets, which must be understood and accepted by your field people, you have no practical way of measuring and monitoring the actual productivity.

This fact is well known in sports and in other industries. You have to set a time limit or a value on an activity in order to be able to measure the performance. If four minutes is the optimum time to run a mile and is generally accepted by runners to be the target, then you have a practical means of measuring the performance of a runner. Once you have accepted four minutes as being an acceptable target, then you can look into the ways and means of achieving it.

If on a given project you have established that your installers should average 100 ft of ¾-in rigid conduit per worker per day as an acceptable target, then you must plan the work to achieve this. You now have a means of measuring the productivity for this operation. Identify the critical operations and target them. The installers should have definite standards against which to measure their performance. The key operations may constitute only a fraction of the job, but by controlling them you will control the whole job. Conduit-installation operations are the ones on which most of the worker-hours are lost. Installation standards for these operations should be set and monitored as the key to your program to increase productivity.

JOB FACTORS AND PEOPLE FACTORS

The actual worker-hours consumed are determined by job factors and people factors. Adverse job factors have been itemized in previous chapters; they generally consist of poor site conditions, poor planning and coordination, insufficient or inadequate instructions, requirements that are too demanding, extreme temperatures, and too many changes and revisions. People factors consist of the attitude and motivation of the workers, their skill and involvement, overtime, shift work, overstaffing, and morale.

Extend Duration of Productive Work Periods

You know all these generalizations. The question is: What can you do in a practical way to improve the productivity in the field? Start with the recognition that in a regular 8-hour day you are probably getting only 3 hours of actual time in which the installers are doing productive work. Therefore, the first step to increase productivity is to expand the amount of time during the day in which the productive work can take place. Your target should be 4½ to 5 hours. In order to accomplish this, you will have to cut down the time consumed by break periods, by material handling, and by nonproductive instructional, laying-out, and informative activities. Refer to Fig. 9.1 to see this in graphic form.

Workers' Shacks

To cut down the time lost during break periods, use collapsible and movable shacks, as shown in Fig. 9.2, that can be moved easily and erected quickly close to the areas where the installers are working. This will cut down the amount of time spent walking to and from the shacks for coffee breaks, lunch, and morning and afternoon clothes change. It will pay you to provide your workers with their own individual containers of drinking water, to cut down the time lost when they go to a

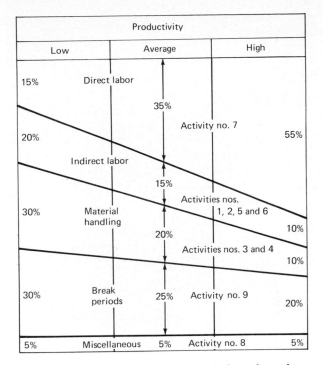

FIGURE 9.1   **Range of percentages of total worker-hours consumed by the various activities.**

central drinking fountain. You have to arrange for practical means of enforcing the contracted period allotted for break periods. If the contract calls for 15 minutes for a break period and 30 minutes for lunch, then you must take steps to ensure that the lengths of these periods are not stretched out.

Typical Workday

By studying Fig. 9.3, you can see that an 8-hour day breaks down into four work periods. You should ensure that the afternoon break does not cut into the last work segment, to the extent that the installers will be reluctant to start a serious work activity because of the short amount of time left before preparation for punching out. This is particularly true on Friday afternoon, when the last segment of this day is probably a dead loss.

The four productive work segments of an 8-hour day can be very seriously eroded by late start, early finish, stretched break periods, extended preparation to start and finish work, material handling, and

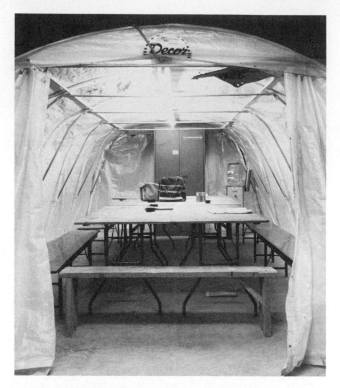

FIGURE 9.2 Movable enclosure used by workers during break periods.

information handling. These actions can make the 8-hour day very inefficient because of the resulting shortness of the four productive work segments. Examine very carefully the logistics of how your field people squeeze out their production in this broken-up 8-hour day. The fact that your industry is suffering low productivity and that substantial schedule and money overruns are common may be partly due to focusing on worker-hours without really controlling or changing the logistic way in which these worker-hours are spent on the job. If there is ever pressure in the industry to decrease the 40-hour week, you should consider four 9-hour days rather than five days of less than 8 hours. The foregoing analysis has shown that the 8-hour day is, in a practical sense, already too short because of the way in which it is broken up.

Material Available for Work

Late delivery of material and delivery of wrong or incomplete material are responsible for many lost worker-hours and low productivity. In-

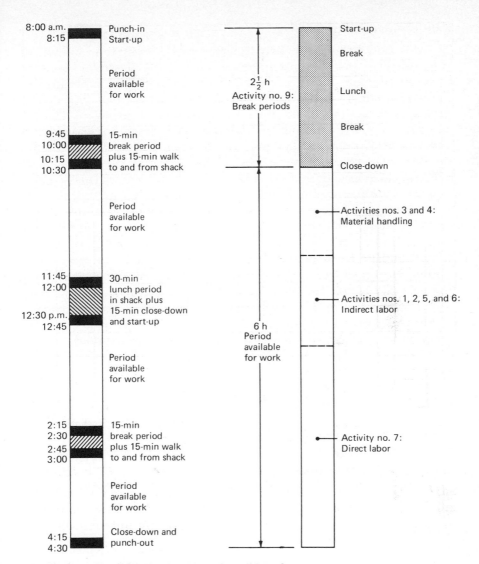

8:00 a.m.   Punch-in
8:15   Start-up

Period
available
for work

$2\frac{1}{2}$ h
Activity no. 9:
Break periods

Start-up

Break

Lunch

Break

9:45   15-min
10:00   break period
10:15   plus 15-min walk
10:30   to and from shack

Close-down

Period
available
for work

Activities nos. 3 and 4:
Material handling

11:45   30-min
12:00   lunch period
in shack plus
12:30 p.m.   15-min close-down
12:45   and start-up

6 h
Period
available
for work

Activities nos. 1, 2, 5, and 6:
Indirect labor

Period
available
for work

2:15   15-min
2:30   break period
2:45   plus 15-min walk
3:00   to and from shack

Activity no. 7:
Direct labor

Period
available
for work

4:15   Close-down and
4:30   punch-out

FIGURE 9.3   Breakdown of a normal working day.

stallers must have material on time in order to do their work. This calls for proper planning and scheduling. It also calls for common sense, to stock accessory materials rather than order small quantities over and over again with a resulting loss of time. Every worker, or at least every team, should be provided with a lock-up box on wheels for material and tools, which can follow them around as they work. Material should be stored as near as possible to the location where the work is taking place.

Role of Architect or Engineer

There is a prevailing misconception that productivity is an individual responsibility—that all contractors must somehow solve their own productivity problems. That belief is only partly valid. The imperative to improve productivity concerns the whole industry, including design professionals, general contractors, and construction managers. The design professionals do not sufficiently relate their design to the very real and practical problems that may result when their design is implemented in the field. Countless worker-hours are lost, very often for no valid reason, when contractors are forced to install material at excessive heights or in inaccessible locations.

Uncoordinated designs, leading to interference between the trades, and continuous changes interfere with project planning and worker morale. Incomplete and poorly prepared drawings and specifications have a similar adverse effect on productivity. Failure to foresee or even understand the practical problems encountered by the contractor in the field and to take these consciously into account in the design is an important reason for low productivity in the construction industry.

For example, on many projects such as hotels and office buildings which require much branch wiring, you will save a tremendous number of worker-hours when the engineers specify BX (flexible armored cable) instead of conduit and wire. These projects usually have false ceilings and stud partitions. The ceilings are obstructed by beams, ventilation ducts, plumbing, and sprinkler piping. Wiring in the stud partitions very often is run vertically, floor to floor, and the sleeves in the floors rarely line up. As a result, running conduit in the ceilings and walls is very difficult and time-consuming. BX wiring is more than adequate; the installation is fast and efficient and lends itself to prefabrication. Labor-saving requirements should be an important consideration in the design.

Teamwork

By the same token, failure by construction managers and general contractors to provide adequate hoisting facilities, material delivery and traffic control, and other site facilities is an important contributing cause of low productivity. There is an inexcusable, penny-wise and pound-foolish attitude in this regard. An electrical contractor will deliver 10-ft-long conduit in conduit wagons to the site and find that the material hoist is too small to allow the conduit wagons to be wheeled onto it. Instead, the conduit must be stacked piece by piece for hoisting, which results in a large increase in the material-handling cost. Countless worker-hours are lost by workers waiting for elevators or hoists because of insufficient facilities on the jobsite. Trucks are tied up waiting for hoists, and worker-hours are lost because of lack of proper ramps for unloading. It is not sufficiently accepted by the various parties on a construction site that they must cooperate and come up with an efficient plan to effectively handle the movement of workers and material as an important step toward improving productivity.

Morale

Dirty toilets, poor lighting, inadequate heating, too few hoists or elevators, arbitrary rules and regulations, and autocratic and abrasive foremen will cause employee frustration and resentment. Resentment will lead to a lowering of morale. So will the impact of many changes. Why rush to install the work if it will be changed tomorrow? Low morale equals low productivity. Napoleon has been quoted as saying that an army's effectiveness depends on its size, training, experience, and morale and morale is worth all the other factors combined. Good morale is the product of planning, teamwork, communication, and cooperation of all the parties on the construction site.

The Two-Worker Team

Old habits die very hard. One of the most pervasive habits is that of the installers' working in teams of two. This very often results in two people doing the work of one installer. One holds the ladder while the other is working. The industry has not focused sufficiently on this inefficient style of work. You should encourage the tool industry to design the types of tools that favor one-worker activities rather than the normal two-worker team activity: for example, a nonskid ladder with a rig to support a fixture while one worker installs and connects it. You should plan your work so that the installers either work alone whenever possible or work in crews where every worker has a productive task to perform.

Overtime

Avoid overtime if at all possible. A short period of overtime is not too bad, but continuous overtime is wasteful and nonproductive. If it is absolutely necessary, alternate one week of overtime with one week of regular hours. Shift work is better than overtime, and six 9-hour days are better than seven 8-hour days. The workers must have time to rest and relax. Table 9.1 shows the percentage loss of productivity due to

TABLE 9.1 **LOSS OF PRODUCTIVITY DUE TO OVERTIME, AS A PERCENTAGE OF TOTAL WORKER-HOURS**

DAYS PER WEEK	LOSS OF PRODUCTIVITY, % OF TOTAL WORKER-HOURS				
	8-HOUR DAY	9-HOUR DAY	10-HOUR DAY	11-HOUR DAY	12-HOUR DAY
5	0	10	15	20	25
6	10	15	20	25	30
7	20	25	30	35	40

overtime. What many people don't realize is that the decrease in productivity due to overtime not only affects the overtime portion, but also spills over into the regular working day. In prolonged overtime, you may even end up getting less production than you would have obtained without overtime. The productivity per worker-hour decreases continuously during overtime because of fatigue, lower morale, absenteeism, spoiled work, and poorer supervision. The owner or general contractor is usually willing to pay for the premium portion of the wages, but you lose out because of the lower productivity throughout all the hours worked during the affected weeks.

Temperature

Temperature affects the productivity of construction workers. The optimum working conditions are taken to be 20°C and 50 percent humidity, with a 3 km/h wind. Conditions above or below these optimum readings will affect the productivity of the work. Obviously, rough work will be affected in a different manner from fine work such as control wiring. Table 9.2 shows the changes in productivity for different temperatures. This table assumes that the workers are properly dressed for the particular temperatures that prevail.

Overstaffing

Overstaffing the project is an inefficient way of making up lost time or gaining time, particularly if the time has been lost because of poor job management or factors beyond the contractor's control. Overstaffing violates the principle of sizing the gang to the amount of work to be done in order to achieve optimum productivity. The size of the crew should be kept to a minimum commensurate with good performance and good planning. Try to assign workers with previous experience in the type of work involved. Overstaffing will result in the installers'

**TABLE 9.2   ESTIMATED EFFECT OF
TEMPERATURE ON WORK EFFICIENCY***

EFFECTIVE TEMPERATURE, °C	LOSS OF EFFICIENCY, % OF TOTAL WORKER-HOURS	
	ROUGH WORK	FINE WORK
40	60	†
30	40	†
25	30	†
20	0	0
5	0	15
0	0	20
−5	0	35
−10	5	50
−20	10	60
−30	25	90
−35	35	100‡

*Standard conditions are taken as temperature, 20°C; humidity, 50%; wind, 3 km/h.

†No estimate available, but loss of efficiency is probably somewhat less than corresponding loss for rough work.

‡Probably no work can be done.

working one on top of the other and being forced to scramble for the same facilities, equipment, and services. It may also result in your having to draw less productive workers from the labor pool and in a dilution of supervisory controls.

Prefabrication

In order to increase productivity on the job, base your style of work on the extensive use of prefabricated assemblies. Irrespective of the size of the job, prefabrication will result in a dramatic increase in productivity. You will probably run into resistance, both active and passive, when you implement this style of work. You will have to be very committed and carry on a determined campaign with your foremen in order to win them over to this modern way of working. Many of them, no doubt, are more comfortable with the old ways that don't require as much planning and thinking ahead. They will be won over when they realize how much easier it will be for them to meet their performance targets when their installers work with prefabricated assemblies. This is particularly true when it applies to repetitive operations.

OPERATION LABOR TARGETS AND MONITORING

The analysis of how to improve productivity on the jobsite has emphasized two absolute requirements:

1. Increase the period in the regular eight-hour day in which the productive work activity takes place to an optimum of five hours.
2. During these five productive work hours, increase the performance of the installers so that they meet your optimum labor targets.

Increased productivity is not necessarily a question of working harder; it is a question of working smarter.

You will have to set up a regular and continuing program to monitor the actual performance of the key operations. You can expect a lot of resistance from your field people. The foremen will complain that they are already overburdened with paper work and this will add to their work load. However, you must win them over to the understanding that you require accurate performance records in order to manage and control the work.

For a control procedure to be effective and for it not to die out in time, it must fit in naturally with the way that the work proceeds and it must not attempt to change human nature. It is difficult to effectively monitor the performance of each individual worker. It is difficult to get back meaningful or accurate data from the foremen. After all, they must live with their workers and they are are probably members of the same union. You, therefore, have a battle on your hands to make your field people understand that you must get back regular productivity records of individual or team performance on the key operations, particularly conduit installation. You require this type of regular productivity check in order to ascertain the reason why an installer is, for example, averaging 50 ft per day of ¾-in rigid conduit when you have estimated 100 ft. Is this the result of job conditions, poor planning, or a poor attitude to the work? Getting this type of on-site performance data will allow you to take the appropriate action to rectify or mitigate a possibly deteriorating situation.

SIZING THE CREW

An ounce of prevention is worth a pound of cure. The name of the game is to prevent the loss of worker-hours in the first place, and not necessarily to fill out elaborate records to tell you where the hours have already been lost. Therefore, along with the performance data for individual installers working on the key operations, the most effective overall control procedure is to target the operations as they are carried out in the field, and to monitor them. If a foreman has to rough in a slab, for example, that foreman receives a target of a given quantity of workers and a given quantity of days by whom and in which the work must be completed. The quantities of workers and days are derived from the quantity of material that is to be installed, the optimum labor units applicable to this material, and the schedule considerations that must be taken into account. This is how an estimator works out the worker-hours, and this is how the foreman must target the field labor for a given operation.

When the field foremen target the labor for work operations in this way, they are relying on a scientific manner of sizing the crews and setting the time limits, rather than on chance or job pressures to dictate what they should do. They are working with realistic labor targets, which each crew is responsible for achieving, and the whole crew will be forced to push the laggards if the targets are to be met. Target an operation by establishing the quantity of workers required and the quantity of days the operation should take. Otherwise, the foreman will allow job conditions to dictate the quantity of workers and days. By setting targets, you control the job; otherwise, the job controls you.

Plan and organize the work around the basic premise that the operation targets must be met. The size of the crews and the overall size of the gang on the project should be firmly tied to the optimum labor units that apply to the quantity of material installed at any given period. Otherwise, your labor costs will be analogous to a ship without a rudder on a stormy sea. You won't know where you will end up or whether, in fact, you will be shipwrecked at the end of the journey. Cost management is a hard taskmaster, and it has been constantly resisted by your people in the field. To obtain better productivity, you should target and monitor the labor performance and control the factors that affect it.

The following example illustrates how the major electrical labor operations on each typical floor of a highrise building were programmed and controlled. The area of the typical floor was about 20,000 ft², and the major operations were identified as follows.

1. Installation of conduit, wire, and accessories for lighting. Conduit assemblies were fabricated in the shop and consisted of lengths of conduit fitted with a box at one end, with the wires pulled in and tagged according to circuit number. Each box on the assembly was fitted with two BX fixture leads, with a fixture-mounting plate installed at the end of each lead. The respective fixture circuit numbers were marked on each plate. The field labor operation, therefore, consisted of installing these prefabricated assemblies on the ceiling of the typical floor.

2. Installation and wiring of receptacles in the ceiling space for future PAC poles (movable power and communication service columns which provide receptacle and telephone outlets in open-landscaped office layouts). The receptacles were specified to be mounted on brackets supported on the ceiling at a distance of 18 in below the ceiling in a modular pattern. These 18-in-long brackets, with a box at the top connected with EMT (electrical metallic tubing) conduit to a box at the bottom containing the receptacle, were prefabricated in the shop. The connecting branch conduit and wiring assemblies were also prefabricated in the shop. The receptacles and connecting branch conduit were identified as to circuit number and location. The field labor operation consisted, therefore, of installing the prefabricated bracket assemblies and the prefabricated branch wire assemblies.

3. Installation of conduit, wiring, and accessories for low-voltage switching.

4. Installation of 2-in EMT conduits and accessories for telephones. All bends and offsets were made in the shop.

5. Installation of conduit, wire, and accessories for the auxiliary systems (fire alarm, door surveillance, security tour, energy monitoring, clocks, emergency communication, and emergency telephones).

The worker-hours targeted for these operations on the typical floor are listed in Table 9.3.

The building schedule called for a typical floor to be roughed in every two weeks. It was established that the fabrication shop could manufacture all the assemblies for a typical floor in a two-week period. This work

**TABLE 9.3   WORKER-HOURS NEEDED FOR MAJOR
ELECTRICAL OPERATIONS ON TYPICAL FLOOR**

OPERATION	WORKER-HOURS TARGET
Installation of conduit, wire, and accessories for lighting	240
Installation and wiring of receptacles in ceiling space	220
Installation of conduit, wire, and accessories for low-voltage switching	160
Installation of EMT, conduits, and accessories for telephones	270
Installation of conduit, wire, and accessories for auxiliary systems	550
Total	1440

was scheduled in the shop so as to get a head start on the field. The size of the site crew to install the conduit work on the typical floor, based on Table 9.3 and in conformity with performance targets, was derived as follows. Each worker consumes 80 worker-hours in a two-week period. Therefore, the quantity of workers required is

$$\frac{1440 \text{ worker-hours}}{80 \text{ worker-hours/worker}} = 18 \text{ workers}$$

The point of this exercise is to emphasize that the sizing of a crew should never be spontaneous or haphazard. It shouldn't just happen. You must decide on the quantity of workers for a given operation, and give them a target quantity of days in which to complete it, based on an analysis similar to the one described.

CONTROL OF SLACK

A program to improve productivity has to deal with the effects of *slack*. Slack is built into the very nature of work on a jobsite and the manner in which it proceeds. It is difficult to control because it is invisible. Suppose you find yourself in a rush program to meet an opening date. The job is crying for more workers, but there are none to be had. Yet somehow the work gets done on time with the same crew. Or else the job tempo slows down, but it takes the same quantity of workers to complete it, even though the work period has been stretched out. What you are getting is an adjustment of the slack time. Only by targeting the labor, so that you control the size of the crews and the maximum quantity of workers on the job, will you be able to control the slack. All the inefficiency, bad habits, and lost time dissolve into slack and become invisible. Instead of reducing the slack, the jobs will call for more workers. A job crew may be built up to service the peak demands of the work program, and during the regular periods the additional worker-hours are absorbed by slack.

How do your foremen presently establish the size of a crew to do a given package of work, or how does your project manager determine the overall size of the gang for the job? How do they determine the length of time that a crew should be engaged on a work package? If you look into these matters, you will see that the crews are built up to satisfy job demands and pressures—and to accommodate the prevailing slack. Only by having a planned work program based on realistic targets and performance standards can you control the job and improve productivity. You don't need more paper work and reports to tell you that you are losing worker-hours. You need a style of work that prevents the loss at the source.

10
Prefabrication

SHOPWORK INCREASES PRODUCTIVITY

You have been brought up to do the work in the field. In these new times, which are characterized by rising labor costs and falling productivity, it is necessary that you do more of the work in your shop. The level of productivity achieved when you fabricate assemblies in your shop is substantially higher than when you do the same work in the field. In the field, your employees work in an environment which is often cluttered and dirty, under less than favorable conditions, and they are subject to many obstructions and distractions. As one example, consider the time lost before lunch periods and break periods, when your workers must gather their tools and lock them up. They must then spend more time retrieving them and setting up when they resume their work.

These constraints do not exist in a shop, where the work is done under factory conditions, with bench-type power and air-operated tools. During the break periods the tools are left in place. The work is organized in an assembly-line manner and is done under the most favorable working conditions. It is not surprising, therefore, that the time consumed in fabricating assemblies on the jobsite can be reduced drastically when this work is done in the shop. As a positive by-product, prefabrication forces your field people to think ahead and plan ahead. You always do better when you are forced to look ahead.

The field is not suited for the economical fabrication of assemblies. Whether you are a small contractor or a large one, you should aim at setting up a fabrication shop if at all possible. You should furnish your shop with the best available production tools and the most skilled workers that you can find. For your shop to be successful, you need a special type of person to head it. The person whom you choose for this management position should be a good organizer, have an inventive mind, be mechanically inclined, and have strong leadership abilities. With a well-equipped shop headed by such a person, your payback will be fast and continuous.

The purpose of having a shop is to produce repetitive-type material assemblies at the lowest possible cost. Yesterday's methods of working are just not good enough. Your planned and coordinated on-site work program should be backed up by a production shop with machines and jigs for volume output.

Consulting engineers have been slow in incorporating standardized assemblies into their designs, as a means of encouraging the wider use of prefabrication by contractors. Consulting engineers, too, have a responsibility to interface their designs with improved field installation procedures, in order to help increase productivity.

Field people often resent prefabrication because they think it diminishes their status and will turn them into installers. The unions, too, often oppose prefabrication because of the mistaken attitude that it will diminish the amount of work available for their members. These are short-sighted attitudes. The need to increase productivity is an absolute imperative for the continued health and growth of your industry. Your industry is not immune from the pressures to reduce costs, in order to make the product sufficiently affordable to attract increasing numbers of builders and buyers.

To make shop fabrication the hub around which your field installation work revolves will call for the utmost dedication on your part. In order to achieve it, you may be forced to withhold certain types of tools from your jobs, so that specific operations and assemblies will have to be done in your shop. This applies to items such as large conduit bends or conduit offsets and bends of any size that are required in quantity. An example of such conduit prefabrication is illustrated in Fig. 10.1. Most contractors have done some form of prefabrication at one time or another, but not too many have made it the focus of their style of work. Do not underestimate the time and effort that you will have to expend to win over your employees to this way of working.

For fabrication to be successful, you have to train your field forces to prepare coordination drawings from which they will extract the assemblies that can be built in your shop. Accurate coordination drawings or sketches are required by your shopworkers in order to build assemblies that will fit in place on the job. If your volume of work is large enough and you are operating a sizable shop, it may be worth your while to assign an experienced draftsman to the shop. This draftsman, by working closely with your field coordination personnel, will become an ex-

FIGURE 10.1   Prefabrication of conduit bends.

pert in laying out the work for prefabrication and in making up sketches for the assemblies required by the various jobs.

The prefabrication style of work will force your field personnel to plan ahead. For example, while making up a shopwork order for conduit bends or offsets, your foreman may become aware of some interference or obstruction to the running of the conduit. There will then be enough lead time that the problem can be rectified before the work starts, rather than be discovered only after the work has started, with disruption and delay resulting. Worker-hours are lost frequently and continually in on-site fabrication because the problems surface only after the work has started, rather than being apprehended and solved beforehand.

You require adequate space for your shop. There is a fair amount of material handling involved and some intermittent storage. When you start up your shop and begin to discover its beneficial impact on your profitability, you may find yourself looking for additional space. Don't skimp on the size of your shop or on the quality of the tools that you use in it. Choose tools that are rugged and that are suited for assembly-line production. Search out workers for your shop who also have mechanical, lathe, and welding experience.

LAYOUT OF A FABRICATION SHOP

The following is a description of the production shop of Electrical Contractor E. The shop layout is shown in Fig. 10.2 and occupies about six

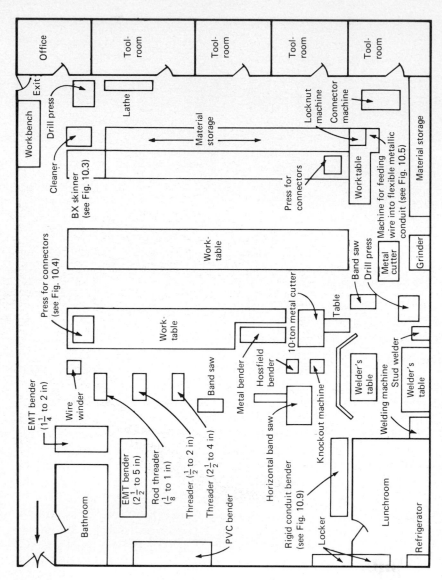

FIGURE 10.2 Layout of a prefabrication shop.

Office

Tool-room

Tool-room

Tool-room

Tool-room

Exit

Workbench

Drill press

Cleaner

BX skinner
(see Fig. 10.3)

Lathe

Material
storage

Material storage

Locknut
machine

Connector
machine

Worktable

Press for
connectors

Machine for feeding
wire into flexible metallic
conduit (see Fig. 10.5)

Press for connectors
(see Fig. 10.4)

Work-
table

Work-
table

Band saw

Table

Band saw

Drill press

Metal
cutter

Grinder

10-ton metal cutter

Metal bender

Hossfield
bender

Knockout machine

Welder's
table

Stud welder

Welding machine

Welder's
table

EMT bender
($1\frac{1}{4}$ to 2 in)

Wire
winder

EMT bender
($2\frac{1}{2}$ to 5 in)

Rod threader
($\frac{1}{8}$ to 1 in)

Threader ($\frac{1}{2}$ to 2 in)

Threader ($2\frac{1}{2}$ to 4 in)

PVC bender

Horizontal band saw

Rigid conduit bender
(see Fig. 10.9)

Locker

Bathroom

Lunchroom

Refrigerator

108

thousand square feet of area. Twenty workers, more or less, are employed year round in the shop; this represents about 10 percent of the contractor's total work force. Many of the jigs, machines, and tools have been modified or made up in the shop in order to speed up the prefabrication operations or to perform them more effectively and efficiently. The shop contains the following tools:

1	2½- to 4-in EMT bender with electric pump
1	½- to 2-in EMT bender with electric pump
1	½- to 2-in rigid conduit bender with electric pump
1	2½- to 5-in rigid conduit bender with electric pump (Fig. 10.9)
1	Bender for PVC conduit
1	Spot welder
2	Welding machines
1	Welding torch set
2	Portable grinders
1	Bench grinder
1	Large horizontal band saw
2	Small band saws
1	½- to 2-in power threader
1	2½- to 4-in power threader
1	⅛- to 1-in rod threader
1	Abrasive saw cutting machine
1	Press with attachment for making one-shot offsets on small-size EMT conduit
1	BX cutting machine (Fig. 10.3)
1	Air compressor
1	Set knockout punch, ½ to 2 in, with electric pump
1	Set knockout punch, 2½ to 4 in, with electric pump
1	10-ton metal cutter
1	Locknut-tightening machine (Fig. 10.4)
3	Punch press machines for fastening connectors to BX or flexible metallic conduit
3	Wire-skinning machines

1   Machine for fishing wires into flexible metallic conduit (Fig. 10.5)

2   Drill presses

       Assorted screw guns, electric drills, and other hand tools

The BX cutting machine was developed in the shop and is illustrated in Fig. 10.3. This machine makes three cuts in seconds, and the BX is then ready for skinning. It can also be used for cutting flexible metallic conduit to given lengths. One worker operates this shop-built rig, which cuts and strips BX cable from a reel after pacing out the cable to a measured stop set up on the worktable. The rig cuts and strips off the armored cable at the proper length, and leaves 6 in of conductor exposed at the end. The next step is to connect the BX lead to a connector or to a box in which connectors have been installed in a previous operation.

FIGURE 10.3   **BX (flexible armored cable) cutting machine.** *(Courtesy of Electrical Construction and Maintenance. Reprinted by permission.)*

A machine for installing connectors onto boxes for fastening to BX or EMT is shown in Fig. 10.4. It can also be used for installing threadless connectors on boxes for fastening to rigid conduit. It is operated by compressed air and can be adjusted to exert exact tightening pressure on the locknut. The boxes are preassembled with the required quantity of connectors. The BX, flex, or conduit is fastened into the connectors

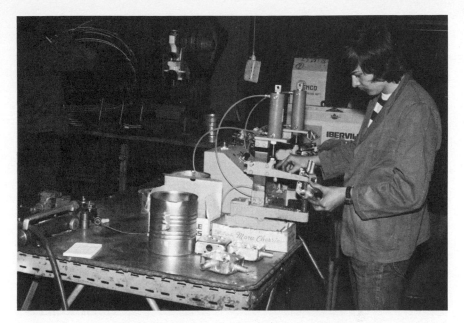

FIGURE 10.4 **Machine for tightening locknuts for connector and box assemblies.** *(Courtesy of Electrical Construction and Maintenance. Reprinted by permission.)*

by using a special punch press. The special connector was designed to eliminate the need for anti-shorts on the BX. The wire is skinned by using a motor-operated wire-skinning machine.

Figure 10.5 shows a machine which automatically fishes two, three, or four no. 14 or no. 12 conductor wires into precut lengths of flexible metallic conduit for the economical fabrication of fixture leads. At one end of the conduit lead is the special connector, and at the other end is a fixture-mounting plate. If, as is often the case, you require 2 ft or more of wire in the fixture, then the loss in stripping more than the standard 6 in from the end of the BX makes the use of BX more expensive than cutting the flexible metallic conduit to the required length and allowing the wire to continue through it for the desired length.

The rig shown in Fig. 10.5 was developed in the shop as part of a bench-assembly operation to prepare fixture leads. In this case, two sets of no. 14 conductors are pulled from reels and are fed under and through the chain-driven drum in the background. After the conductors are guided to the opening, they are pushed into 6-ft lengths of empty flexible conduit. A pivoting knife cuts the conductors at the proper length. Up to 10-ft lengths of EMT can also be prewired at this bench.

The prefabrication of boxes that will be installed in concrete is illus-

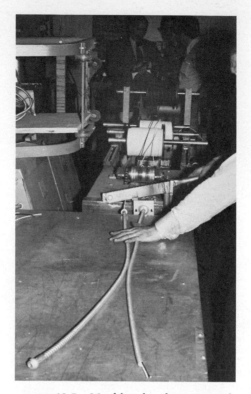

FIGURE 10.5 Machine for the automatic feeding of wire into precut flexible metallic conduit. *(Courtesy of Electrical Construction and Maintenance. Reprinted by permission.)*

trated in Fig. 10.6. These conduit boxes, with rigid threadless connectors attached, are covered with plastic bags. Threadless connectors are used because they will save many hours that would otherwise be spent to thread the conduit. In the case of modular construction, the conduit can be precut and wired in the shop.

Prefabricated BX harnesses loaded into mobile wire baskets for shipment to the jobsite are shown in Fig. 10.7. The baskets are fitted with wheels which allow them to be moved along with the installation crew. These wire-mesh-sided storage boxes can carry all types of materials, shipped from your shop or from your jobsite stores, thus cutting down on material handling. The baskets can be handled by forklift or pallet lift and can be stored one on top of the other. They can also be used as fixture transporters and as small reel carriers when the job entails pulling wire.

FIGURE 10.6 **Prefabrication of box assemblies for installation in concrete.** *(Courtesy of Electrical Construction and Maintenance. Reprinted by permission.)*

TYPES OF ASSEMBLIES SUITABLE FOR PREFABRICATION

The above descriptions underline the fact that many repetitive, time-consuming field activities, such as cutting BX and flexible metallic conduit, skinning wires, installing and fastening connectors, and tightening the locknuts when connectors are installed in boxes, can all be mechanized and performed more efficiently and quickly by using machines and jigs in your shop. The machines and jigs are modified or designed to allow your workers to perform the required activities most effectively, and the work is set up in an assembly-line manner for maximum efficiency.

This principle is best illustrated when it is applied to the bending of conduit. On the jobsite, poor working conditions, lack of proper maintenance or parts for your benders, and sometimes lack of expertise can make the bending of conduit very time-consuming and expensive. In your shop you can afford to invest in the best available benders since

FIGURE 10.7 **BX harnesses being loaded into a material-handling basket.** *(Courtesy of Electrical Construction and Maintenance. Reprinted by permission.)*

you are servicing many jobs. You can modify them, maintain them properly, arrange them for most efficient performance, and assign operators who will become proficient and expert at bending conduit.

When you insist on having the major conduit bending done in your shop, you will be under tremendous pressure from your field people to reverse this procedure. Their resistance will be stronger if you withhold the large benders from the jobs, to force those in the field to use the shop for this purpose. They will give you many reasons why you should send benders to the job. If you give in, there will be very little bending done in your shop.

Bench bending equipment for bending EMT conduit with diameters of up to 4 in is illustrated in Fig. 10.8. Most of this type of bending can be more efficiently done in the shop. When bends are made, they are tagged and loaded on a conduit wagon for shipment directly to the point of installation at the jobsite. This reduces the handling time. The conduit wagon shown in Fig. 10.8 is a wheeled rack made from conduit bent into yokes to form a cradle that allows clearance for carrying

FIGURE 10.8   Bench bending equipment for EMT conduit. *(Courtesy of Electrical Construction and Maintenance. Reprinted by permission.)*

bends, sweeps, and offsets. This wagon or dolly is used within the shop during bending operations and, subsequently, for shipping the bent conduit to the job. You can see in the background the table jig for the hydraulic bender, which has permanent braces or stops that speed the horizontal bending activity at workbench height. Extra shoes are stored nearby.

The vertical bending machine for bending 1¼- to 5-in rigid conduit is illustrated in Fig. 10.9. The advantage of vertical bending is that the work can be done standing up. A magnetic protractor or gauge inclinometer near the top of the conduit allows the operator to achieve the correct degree of bend. The machine has a mobile cradle which moves the conduit along the length of the bench as the operator turns a handle. An experienced operator can fabricate a large-size conduit bend in 10 minutes or less. There is no question that you will achieve substantial savings when the conduit, particularly of larger sizes, is bent in your shop rather than by the installer in the field. A conduit-bending specialist can handle most of the bends required by your jobs in the area in which you operate, by using such an electric hydraulic bender mounted on a rig that puts all the conduit at optimum working height to ensure accuracy and speed. The mechanic makes up the bends by reading the job production sheets, which show the details of the bends required.

The size of the production shop described above is obviously not within the means of many contractors. However, the concept and many

FIGURE 10.9   Upright bender for rigid conduit. *(Courtesy of Electrical Construction and Maintenance. Reprinted by permission.)*

of its features certainly are. Your shop and its size are determined by the type and volume of the work that you do. The size of the staff and the quantity of workers there will vary with your work load. In that respect, the shop is no different from any of your jobs. The quantity of workers bears a direct relationship to the work on hand, and the workers are drawn from or released to your active jobs or the industry labor pool. Obviously, you will try to hold on to and protect the more talented and experienced workers. It is probably a good idea to move workers out of the shop and into the field occasionally to prevent fatigue.

PRODUCTION SHEETS AND SHOP REQUISITIONS

The concept of using factory-style assembly-line methods to improve the productivity of your jobs is indispensable. Every repetitive operation on your jobsite should be reviewed for possible prefabrication in your shop. You will probably find, depending on the type of work in which you specialize, that many assemblies, which you can standardize, keep recurring. When this is the case, arrange to make up preprinted production sheets of these standard assemblies, which your jobs can use for ordering from the shop. The following are examples of some standard assemblies:

- Bends and offsets (Fig. 10.10)
- Angle iron, channel, unistrut, and flat iron supports (Fig. 10.11)
- Angle iron and channel brackets (Fig. 10.12)
- Box assemblies and pigtail leads for BX and flexible metallic conduits (Fig. 10.13)
- Feeders, parallel-reeled and cut to length (Fig. 10.14)

The foreman fills out the production sheets by listing the exact measurements and quantities required and covers them with a shop requisition (Fig. 10.15). Only the top portion of the shop requisition is filled in. The foreman signs the shop requisition, which is a five-copy set, and retains the third copy. The rest are sent to the shop.

When the order is completed, the shop lists the materials and the worker-hours consumed by the order on the portion of the shop requisition provided for this purpose. The first and fourth copies of the shop requisition accompany the completed assemblies to the job, where the fourth copy is signed and returned to the shop as the signature copy. The top shipping copy is retained by the job. The second copy is sent to the costing department, and the fifth, or production, copy is retained by the shop. Thus the costs of the materials and labor expended by the shop for the various jobs are assigned to the respective jobs for cost recording and analysis.

Aside from the preprinted production sheets, which you can have made up for the particular items that recur frequently in your particular type of work, the jobs make up production sheets for any type of assemblies which they require for their installation. Figure 10.16 is an example of a production sheet for a highrise office building installation, where the branch wiring was done in BX and was completely prefabricated in the shop. The assemblies consisted of strings of BX connected to boxes, with the fixture leads connected to them. Caddy clips were installed on the back of the boxes. The strings of boxes were tagged as to floor, section, and circuit numbers, as called for in the respective production sheets. The completed assemblies were placed in wire baskets and shipped directly to the area in the project where the assemblies were installed. The completed assemblies were installed by fasten-

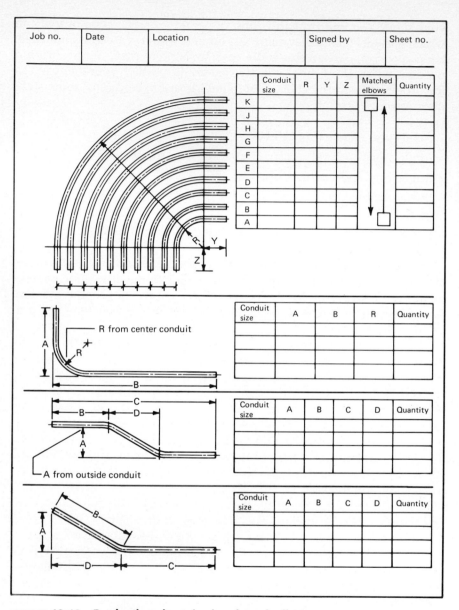

FIGURE 10.10 Production sheet for bends and offsets.

Job name	_____	Production sheet no. _____
Job no.	_____	Sheet no. _____ of _____
Location	_____	JR no. _____
Date ordered	_____	Date required _____
Ordered by	_____	

Unistrut channel — Figure 1
Steel channel — Figure 2
Angle iron — Figure 3
Flat iron — Figure 4

Quantity	Figure	A	B	C	Catalog no.	Gauge	Size	Face 1			Face 2			Face 3			Paint
								X	Z	Dia.	X	Z	Dia.	X	Z	Dia.	

Remarks

Quantity	A	B	C	Dia.	Full thread	Paint	Remarks

FIGURE 10.11 Production sheet for angle iron, channel, unistrut, and flat iron supports.

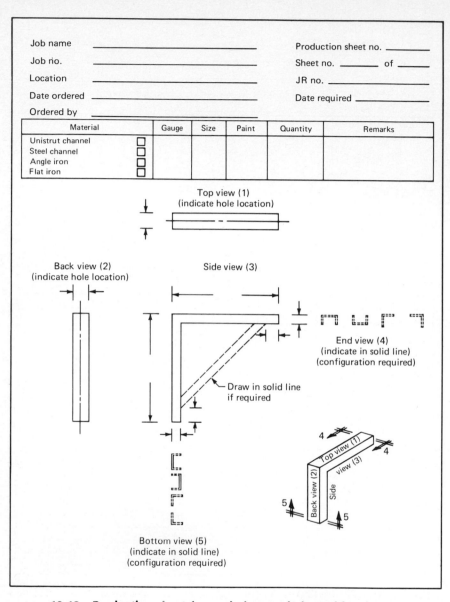

FIGURE 10.12 Production sheet for angle iron and channel brackets.

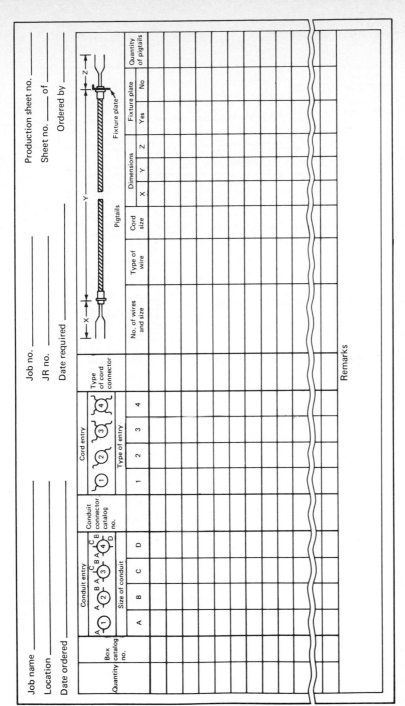

FIGURE 10.13 Production sheet for box assemblies and BX or pigtail leads.

121

Job name _____			Job no. _____			Production sheet no. _____				
Location _____			JR no. _____			Sheet no. _____ of _____				
Date ordered _____			Date required _____			Ordered by _____				

Item no.	Reel tag	Feeder no.	From	To	No. of wire	Size of wire	Type of wire	Length and color				Total length	Remarks
								Red	Black	Blue	White		

FIGURE 10.14   Production sheet for parallel reeling and cutting to length of feeders.

ing the boxes with their caddy clips to pencil rods which had been previously installed to support the BX wiring.

It is evident that this style of work is much more efficient and productive than the old style of cutting and assembling the materials on the job prior to installation. The production of these assemblies in the shop demanded more thorough planning of the most efficient way to install them on the job. The time consumed to install them was a fraction of what it would have taken had the work been done in the old way.

HANDLING OF PREFABRICATED ASSEMBLIES

The cost pressures of these new times demand that as many of the repetitive assemblies as possible should be fabricated in the shop or under shop conditions. The role of the field is to plan, coordinate, and install these efficiently prefabricated assemblies. In your discussions with the engineers and designers who are associated with your work, you should champion this concept and encourage them to take prefabrication into account in their designs, and also to incorporate better material-handling and easier fastening methods therein.

All materials and assemblies are moved around the shop or to the jobsite in wire baskets on wheels, on conduit wagons, or in other types

Job name			Job no.	
☐ Contract			Fab. no.	
☐ CO no.			Date	
☐ WO no.				
Quantity	Description		Date required	Sketch no.
Quantity	Material		PO no.	Supplier

Classification	Regular hours	Double time	Total hours	Rate	Amount
Shop labor					

Received above merchandise in good condition	No. of pieces	Ordered by

FIGURE 10.15 **Shop requisition form.**

of movable containers. These are readily loaded onto a truck by using a forklift or pallet lift, and once unloaded, they are easily moved to the point of use. Wire baskets loaded with BX assemblies are illustrated in Fig. 10.17. Another example is shown in Fig. 10.18, in which prefabricated conduits, prewired, with boxes and BX leads attached, have been loaded on a conduit wagon and are being moved by a worker to the installation area.

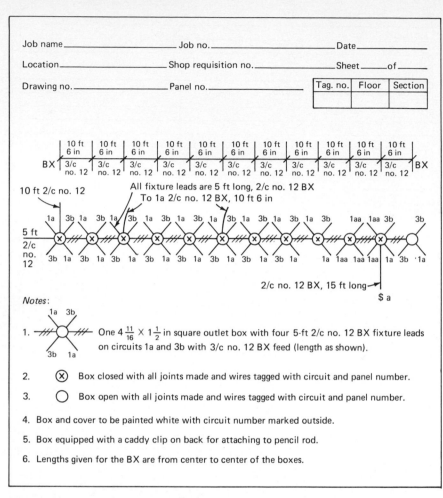

Job name_____ Job no._____ Date_____

Location_____ Shop requisition no._____ Sheet____of____

Drawing no._____ Panel no._____

Tag. no.	Floor	Section

FIGURE 10.16 **Field production sketch to prefabricate lighting branch wiring assemblies.**

TAGGING AND IDENTIFICATION

The keys to the efficient and effective use of prefabrication are proper and precise tagging and identification of the assemblies. The tags and identification should be such that they will not be lost in shipping and handling. They should be clear, so that the correct assemblies will arrive at the right place at the right time. The tagging should identify what the assemblies are, to what project they are going, and where exactly in that project they will be installed. Paint the covers of boxes with designated colors to identify the system and the voltage. In the case of prewired assemblies, identify the circuit numbers clearly.

FIGURE 10.17   BX assemblies in baskets awaiting shipment to a job.

FIGURE 10.18   Prefabricated conduit assemblies in a conduit wagon being wheeled to point of installation.

Your purpose should be to cut out double handling and the time lost in sorting out and trying to identify what goes where. Don't look for shortcuts in this regard. Assemblies for many floors or different areas should not be put in the same basket. Load them and tag them so that assemblies for a given work area are in one or more containers specifically assigned to that area. Arrange your shipments to cut out possible mixups and extra handling, all of which will inevitably result in delays and lost time. You have to educate your field people to be very clear and precise when ordering the assemblies. This will ensure that the assemblies will be properly fabricated to suit the job requirements and will be delivered at the required time to the required location. The field people must understand that the shop is like a factory. Production must be scheduled and programmed to fit the work load, and precise information is the best guarantee that the assemblies will be built to suit the job requirements. Lack of precision in ordering and tagging will defeat the system. Proper planning, ordering, and tagging will pay great dividends in increased efficiency, productivity, and morale. It won't happen by itself, but it is worth the effort.

11
Purchasing

PURCHASING PROCEDURES TO IMPROVE PRODUCTIVITY

Most purchasing procedures develop spontaneously and become habitual. On the way, no doubt, you have picked up your share of bad habits that result in worker-hour losses in the field. You should review your purchasing procedures regularly and bring them in line with the needs of your jobs. When placing your material orders, keep in mind that it is the installed cost and not the first cost that will determine whether you have made a good deal.

Make your purchasing system as simple and direct as possible. All major equipment and large-volume materials are ordered by the purchasing department from reputable suppliers who undertake to respect your purchase conditions and delivery requirements. Set targets for the various packages of material which are to be purchased, so that your purchaser will have a goal to aim at. The purchase order should spell out when you wish to receive the shop drawings and when you require the order to be delivered to the site. Slow delivery, late delivery, and wrong delivery are major impact factors that affect your jobs and cause disruptions and worker-hour losses.

Lines of communication should be straight, direct, and consistent. The primary purpose in placing the purchase order should be to meet the intent of the plans and specifications at the lowest possible price, and also to ensure that you have incorporated the necessary conditions which will help the work on the site to be done in the most practical and efficient manner possible. This calls for clear and consistent communication between the purchaser, the estimator, the contract manager, and the project manager, all of whom are part of the purchasing system. A clear line of communication should also be established between the field project manager and the supplier-manufacturer to ensure that site delivery and installation requirements are taken into account.

Accuracy and consistency must be built into the purchasing system and made routine. A purchase order is a contract and should be made up precisely to describe not only what you expect to receive, but also how and when you require it. All the parties in your purchasing system, including the field, should have updated binders containing the latest catalogs of the equipment and materials which are being installed on the jobsites. It is a good idea to prepare a catalog of common items that are stocked in your warehouse, so that the jobs can easily and precisely order the accessory and fastening materials that fit their requirements.

Arrange to eliminate overlapping and fuzzy operations. The field should receive copies of the purchase orders so that they don't requisition items already ordered. This is particularly true with regard to the handling of back orders. Every link in your purchasing chain should be precisely aware of its function in the initiation of purchases, so as to cut out duplications and delays, as when one party thinks that another is responsible for a particular function.

Establish logical priorities, so that first things get done first and the style of work is methodical and free of panic. It is impossible to do everything at once. Therefore, there must be priorities. The job schedule and job requirements should dictate the scheduling of the purchasing activity.

The various people involved in the purchasing system should study and review the procedures that involve them. In this way they will isolate the particular problems that affect them and find satisfactory solutions. When you take the time to sort through your past experiences and problems, you will be in a good position to establish the most effective purchasing system to suit your operation. The pressure of the escalating cost and complexity of doing work in your industry forces you to review your procedures and to modernize your thinking and method of operation in order to survive.

What is absolutely required is that your project managers interface closely with your purchasing department and also with the suppliers and manufacturers. The principle of incorporating jobsite requirements into the purchase order is of utmost importance. It is not good enough for the purchasing department and the supplier to put all the emphasis on arriving at a mutually agreeable price, whereupon each feels the matter ends. For pennies saved, dollars are lost in the field. There must be a closer and more direct link between the installer and the manufacturer, so that the jobsite coordination details and schedules become a vital part of the purchase order and are understood by all concerned.

The following items are too often neglected by the purchasing department:

1. The vital need to incorporate requirements such as tagging, method of shipping, field coordination details, required delivery dates, shop drawings, and maintenance manuals into the purchase order
2. The advantage of replacing a large number of small purchase orders, very often for the same types of materials, with bulk orders to be released as required by the job
3. Methods to improve the manner of dealing with back orders
4. Methods to simplify and improve the reconciling of a distributor's invoice with the covering receiving report and purchase order, for release of payment

Each purchase order (PO) is a contract. It should clearly define the scope and intent of the order. It should establish the following:

1. The manufacturer and distributor.
2. The total price, or unit prices and discounts.
3. The applicable plans and specifications to which the PO is subject, and the required shop drawings and manuals.
4. The lead time after approval for manufacture and delivery to the site. The delivery arrangement is of critical importance and should be spelled out.
5. The obligation of the supplier and manufacturer to cooperate with the field organization to incorporate job coordination details, marking, identification, shipping arrangements, and delivery requirements. This will expedite the placing of orders, since it separates the purchasing function from the coordination and scheduling function, which should be the responsibility of the project staff.

The purchasing department should *bulk-order* the standard, repetitive types of materials required by the job, as much as is possible and practical. The quantity of purchase orders and job requisitions should be reduced to a minimum. This will require thought and planning to cut down the 4:00 p.m. phone calls for material to be delivered the next morning. It is important to learn from experience by examining past jobs. How many purchase orders were there? How many small ones, and how many per week? How many similar items were ordered many times on different purchase orders when they could have been grouped? This type of analysis pertaining to a large engineering project is shown in Table 11.1.

It is evident from Table 11.1 that most of the purchase orders covered accessory items such as hangers, outlet boxes, plates, fuses, channel, rods, angle iron, anchors, hardware, plywood, bits, taps, fishwire, wire lubricant, acetylene, and propane. The accessory materials, which added up to about 20 percent of the total cost of materials, took up about 70 percent of the purchase orders. When you consider the time and effort involved in making up these small orders, checking them, receiving them, or chasing around for them, you can see how inefficient this style of work can be if the job is allowed to develop spontaneously without adequate planning. How many times have you been at a supplier's order desk and seen electricians lining up to get shelf materials, while their helpers sit in the company trucks with the engines running, waiting for them? Dollars are spent to pick up penny materials.

To accommodate modern data-processing costing systems, materials must be requisitioned and ordered in accordance with designated categories and codes in order to simplify the posting. The designated material codes must be taken into account and will affect the ways in which the jobs requisition material, the warehouse prepares their shipping slips, and the purchasing department makes out the purchase orders.

**TABLE 11.1 PURCHASE ORDERS FOR
A LARGE ENGINEERING PROJECT**

TYPE OF MATERIAL	TOTAL COST, DOLLARS	QUANTITY OF PURCHASE ORDERS
Distribution equipment	475,000	20
Auxiliary equipment	130,000	26
Raceways	200,000	26
Fluorescent fixtures	50,000	1
Incandescent and mercury fixtures	130,000	11
Lamps and fixture accessories	20,000	8
Conduit and wire accessories	250,000	186
Finishing materials	5,000	17
Nonelectrical accessory materials	115,000	234
Conduit	100,000	10
Code wire	50,000	25
Special cable	225,000	15
Rentals	130,000	13

In the case of materials such as conduit, wire, boxes, fittings, and bus duct, the exact sizes, lengths, quantities, bends, and elbows flow from the coordination details and jobsite conditions. The overall size and value of this type of order—that is, the *bulk* of the order—is established from the estimate. This purchase order should contain the general conditions of the order and a listing of the respective *unit prices* and *discounts*. The actual required sizes, fittings, and quantities are released by the job. Otherwise, the actual job releases would so modify a purchase order based on sizes and quantities as to make it very difficult to reconcile the supplier's invoices with the purchase order. The unit price and *release method* will simplify the procedure, since each invoice should match the respective release order.

In the case of switchgear and equipment, the purchase order is usually for a firm price based on a defined description and listing of material. Any revisions or changes to this type of purchase order should be made in writing on a PO bulletin. PO bulletins are numbered consecutively. At regular intervals, purchasing will accumulate all bulletins for a given purchase order, reconcile the total change, and issue an amended purchase order. In this way there will be a written record of all the changes, in order to reconcile the final value of the purchase order.

REVIEW OF THE MATERIALS IN THE ESTIMATE

When your tender is successful, the material in the estimate should be examined in a deeper, more thorough manner, so that the quantities are rechecked and the actual job requirements are taken into full consideration. Material is grouped into logical and designated categories, identified clearly as to quantity and description, and tailored to fit the actual installation requirements. This results in efficient purchasing and will cut down time lost on the job when wrong material arrives or when the delivered material must be modified to fit the job conditions.

1. When ordering items such as wire and cable, take advantage of optimum reel sizes, color coding, parallel reeling, pulling eyes, and tagging, and don't forget the accessories, such as connectors and pulling compound. In the case of larger cable, the reel sizes should be a multiple of a given number of feeder runs and should be as large as can be conveniently handled in order to cut down wastage. Color coding will save worker-hours when it comes to identifying the phases. Parallel reeling will save set-up time and pulling time.

2. Check the major equipment such as switchgear, so that qualifying notes on drawings, in the specifications, or in addenda are ac-

counted for. The purchase order should concern itself with the method of service entry, the optimum shipping size, special lifting accessories, whether the gear is metal-clad or metal-enclosed, the interrupting rating, frame sizes of the circuit breakers, knockouts, and types of fuses. The required features must be incorporated in the manufacturing stage, and the problems should be solved at that time, not in the field, where they will result in delays or additional labor.

3. Check out the transformers for rating, impedance, taps, sound rating, size of connectors or lugs, flexible connectors, vibration isolators, and fan cooling.

4. Identify the panels by a logical system of tagging. Incorporate information such as the point of entry, top or bottom, whether the panels are flush or surface-mounted, the size and location of knockouts, the rating of the circuit breakers, and lug sizes. Every problem solved in the factory, every feature that can be incorporated in the factory that will make the installation easier, will save time and money on the jobsite.

5. Establish whether starters and contactors are mechanically or electrically held, the coil voltage, the number of auxiliary contacts required, and whether any other auxiliary components are called for.

6. Study the type and location of the fixtures and establish a tagging system for easy identification. Identify installation requirements, such as inverted tee bar, recessed with plaster frame, or surface-mounted. Specify your desired shipping method, such as the sizes and types of pallets. By incorporating installation and handling details in the purchase order, you will make a tremendous difference in your labor costs, particularly in repetitive operations such as fixture installations. Every good idea and every saving gets repeated hundreds, and sometimes thousands, of times.

The review of the estimated materials has, as an important purpose, the incorporation in the manufacturing stage of many items that would otherwise have to be done on site under more costly and time-consuming conditions. Any operation that can be done in the factory will be more economical and more efficient than if it has to be done on site. To achieve this you must involve your project people. You have to do this without slowing down the purchasing procedure, which could cause delays in delivery of materials to the jobs. Serious delays in delivery could cancel out many of the advantages gained by good coordination. You can achieve a satisfactory tempo of purchasing by including in your purchase orders an agreement by the supplier-manufacturer that they will meet directly with your project personnel to work out the required coordination and delivery details. This will allow the purchase order to be made up without waiting until all the coordination and field requirements are worked out. The manufacturers must get closer to the jobs if they are to satisfy the productive requirements of the jobs.

PREVENTABLE PROBLEMS

Many problems in the field recur time and time again because the purchasing department or the project management fails to draw on its experience and take preventive action when it counts. The following is an example of a check list of preventable problems that every contractor should accumulate for reference. During the purchasing stage and the shop-drawings stage, check and make sure that the following items as they apply have been taken into account.

Incoming Service Cable

1. The proper size and type of cable have been ordered.

2. The correct jacket, colors, tagging, or marking have been included.

3. Lugs and terminations have been ordered, and termination space has been allowed for stress cones.

Main Switchgear

1. Plans and specifications have been checked to ensure that there are no conflicts, errors, or contradictions.

2. Items shown on plans but not in specifications (or vice versa) are accounted for.

3. The supplier is basing the order on the latest information and addenda.

4. The supplier has included the required protection equipment, specified or required by applicable codes.

5. The equipment is approved by the power supply authority, and their requirements for metering have been accounted for.

6. The equipment which your supplier proposes, which meets his manufacturing standards, is acceptable to and approved by the en-

gineer and the power authority. This is also true for items such as the type of construction, protection, and controls.

7. Items such as flow indications, signs, and nameplates, as called for, are included.

8. The location of incoming and outgoing cables and bus duct, and their sizes and type of terminations, are identified and accounted for.

9. The configuration of the gear and the positioning of the sections are to your best advantage, and it is agreed that they will be delivered in manageable sections for hoisting and handling on the site.

10. There are sufficient removable sections and adequate working space, so as not to interfere with your wiring and connections.

11. The rupturing capacity and UL rating are in accordance with specifications and power authority requirements.

12. Your supplier is agreeable to your required delivery dates, particularly for gear required for heating, testing, and other deadlines.

13. The manufacturer accepts the scope of the testing that may be required and will supply the required coordination curves and relay settings so as not to delay the energizing of the equipment.

14. The supplier is aware of requirements to interface with the equipment of other suppliers, so as to eliminate gray areas and loose ends.

15. The size of the switchgear does not interfere with other neighboring systems. Interference by such long-delivery equipment can result in serious delays.

16. You have thoroughly reviewed and checked the shop drawings before sending them to the engineers for approval. You should make up a dated punch list of questions and answers that have been brought up by you, your supplier, or the engineer.

Secondary Switchboards

1. The low-voltage switchboards have received the same treatment as the main high-voltage switchgear described above.

2. The proper flange details for bus duct are provided.

3. All lug sizes and lugs are accounted for. You have to establish who supplies the flexible connections, if these are required.

4. Arrange to check the switchboard thoroughly and immediately upon receipt for possible damages. Report all damages immediately to the transport company, since they are responsible for damage to the equipment while in transit.

Emergency Generator Sets

1. The engine meets the requirements of the specifications as to overload conditions and the amount of time required to run with the specified percentage of overload. All engine manufacturers have their own standards, and most engineers will write their specifications around a specific manufacturer.

2. You have established who supplies the exhaust system, cooling system, and fuel system.

3. You have verified that the cooling ventilation is adequate, that the engine radiator has the proper flanges to connect to the cooling system, and that the oil-coolant and ventilation systems check out before the engine is tested.

4. You have checked whether the muffler requires a drain pipe to remove water caused by condensation.

5. You have checked that vibration isolators have been ordered in plenty of time and that the resulting alignment of the engine fits in with the connecting ventilation and other systems.

6. Adequate openings have been provided to receive the equipment.

7. The rating and capacity of the generator is as specified, and the respective control panels are accounted for.

Motor Control Centers

1. The number of sections are as called for, including spare sections, and the motor control center (MCC) is of the class specified.

2. The starters are so placed as to cut to a minimum the run between the starter and its respective motor.

3. The motor characteristics have been checked and the proper relays and fuses have been ordered, with the type and time-delay requirements specified.

4. The starters are properly identified, and it has been established whether the motor control center is to be top- or bottom-connected.

5. The manufacturer knows how the MCC is fed, whether by cable, conduit and wire, or bus duct, and the correct-size lugs are included.

6. The supplier knows the type and size of cables, particularly for the large starters.

7. The type of control voltage is established, whether from one master transformer or from individual control transformers in each starter.

8. You have established the number of auxiliary contacts for control and the type of local control for each starter, such as hand-off-automatic (HOA), start-stop, pilot light, and pilot-light-test pushbutton.

9. The MCC will be shipped in manageable sections.

10. The paint finish meets the specifications.

11. The required shop drawings and manuals will be provided.

12. The supplier will meet the required delivery date.

13. The required openings will be left to receive this equipment.

14. Channels are provided for the base, if required.

15. Hoisting equipment has been arranged for and coordinated with the delivery to prevent delays and demurrage charges.

Fixtures

1. You have worked out the methods of shipping, receiving, and storage of fixtures. You have made such arrangements for shipping them on pallets, in baskets, or in wagons, depending on job conditions, as will cut down on storage and handling. You have arranged for the fixtures to be stored close to their point of installation and for them to be delivered in coordination with your installation program, so they don't have to be stored for long periods. You have taken into account that the method of storage should consider security, safety, and possible fire hazard. A cigarette carelessly tossed between cartons can start a serious fire.

2. The fixtures will be properly packaged in order to avoid damage and the consequent delay and expense of obtaining replacements.

3. The fixtures will be checked immediately upon receipt to ensure that they comply with the purchase order and approved shop drawings. You have instructed your personnel not to wait until they are ready to install them, in order to prevent possible delays and additional costs.

4. You have ordered the respective lamps of the specified characteristics, with attention to details, such as color and whether mercury lamps are base up or base down.

5. You have ordered fixture accessories, such as pigtails and removable connection plates, to be factory-installed.

6. You have arranged for a ceiling mock-up and observation samples, where possible, to determine what the problems are before the fixtures will be released for manufacture.

7. You are keeping a running account of quantities ordered, quantities received, and quantities back-ordered. You are keeping track of changes, and canceling fixtures that are changed or eliminated. You are keeping track of, and following up on, back orders since suppliers have been known to mix up orders.

8. You have ordered a percentage of additional ballasts, since a small quantity of ballasts will surely burn out, and time is lost waiting for replacements.

PURCHASE ORDER CHECK LIST

A purchase order check list for switchgear and special equipment is filled out by the estimator or contract manager and is incorporated into the purchase order for this equipment. The check list shown in Fig. 11.1 identifies the critical requirements of the plans and specifications relating to this equipment, and also the jobsite handling requirements. This type of equipment is long-delivery, and errors in ordering will result in costly delays or improvisations to make it conform to plans and specifications.

MATERIAL RELEASES

An important key to improving job productivity is the material release procedure. This basic management function will call for more planning and scheduling on the part of the project staff. In particular, they will have to be aware of lead times required to produce and approve shop drawings and of subsequent manufacturing time so that the equipment will be released, manufactured, and delivered to the site in accordance

HIGH-VOLTAGE SWITCHGEAR

Switchboard No. _____

Plan nos. _____

Specifications _____

Number of sections specified _____

Paint finish and color _____

Service entrance top or bottom _____

Terminators—stress cones _____

Type of cable terminations _____

Power company metering—in switchboard or remote _____

Interlocks _____

Cover panels to be locked/hinged/bolted _____

Identification and flow indicator _____

Type of fuses _____

Test requirements _____

Transformer connections:

    Cable _____

    Throat _____

    Flexible leads _____

Control interconnect with other equipment _____

No. of sections for shipping _____

Operation and maintenance manuals

FIGURE 11.1 Purchase order check list.

with the required project schedules. For the release procedure to be effective, these steps should be followed:

1. The methods of shipping and tagging, the quantities required, and all other shipping instructions must be discussed with the production people of the manufacturer, and their agreement is mandatory. The methods of releasing the equipment or material must be mutually agreeable; otherwise, you will go to a lot of effort to make up and schedule your releases, but the manufacturer's production people may not have been informed or they may have no intention of following your releases.

2. A release should be sent in time for the manufacturer to commence production. If the releases are sent out too early, they may be filed away or forgotten, or you may be confronted with job changes that

HIGH-VOLTAGE TRANSFORMERS

Transformer no. _____

Plan nos. _____

Specifications _____

Paint finish and color _____

Type of taps _____

Type of termination each side:

    Box for cable and stress cones _____

    Box for bus duct:

        Top connection _____

        Bottom connection _____

        Side connection _____

    Throat connection:

        Is adaptor box required to clear cooling vanes? _____

Cable data—lug sizes _____

Flexible connections _____

Pan required under transformer _____

Cooling fans:

    Supplied _____

    Mounts only _____

    Remote connected _____

    Local connected _____

Coolant:

    In transformer _____

    Shipped separately _____

Testing requirements _____

Interconnection shop drawings _____

Specified sound rating _____

Right- or left-connected _____

Any special positioning of cooling vanes _____

Operation and maintenance manuals _____

Shipping requirements _____

FIGURE 11.1   Purchase order check list *(continued).*

require revision of the releases. Therefore, the way to schedule a
release is to look backwards. Establish the date on which you want
the particular released material to arrive at the jobsite. Calculate the
number of days or weeks required to manufacture and ship, including a factor of safety for the inevitable delays and broken promises,

LOW-VOLTAGE SWITCHBOARDS

Switchboard no. _____

Plan nos. _____

Specifications _____

Number of sections specified _____

Paint finish and color _____

Interlocks _____

Type of circuit breakers _____

Lift-out mechanism required _____

Spaces or spares:

   Equipped for breakers _____

   Empty _____

Entrance:

   Top _____

   Bottom _____

   Bus duct _____

   Cable _____

   Throat to transformer _____

   Flexible leads _____

Lugs—Single or Double _____

Lug size _____

Side panels:

   Hinged _____

   Bolted _____

Fuses:

   Type _____

   Spares _____

Test requirements _____

Spare parts _____

Identification and flow indicator _____

Shipping requirements _____

Hoisting requirements _____

FIGURE 11.1   Purchase order check list *(continued).*

PANELS

Panel type _____

Plan nos. _____

Specifications _____

Paint finish and color _____

Lug type _____

Lug size _____

Lugs:

   Single _____

   Double _____

Entrance:

   Top _____

   Bottom _____

Surface-mounted _____

Flush-mounted _____

Special locks _____

Fuse type _____

Breaker type _____

Main breaker required _____

Low-voltage control section includes terminals and wiring _____

Identification:

   Lamicoid _____

   Color _____

FIGURE 11.1   Purchase order check list *(continued).*

and subtract this time from the desired delivery date. This will establish the release date.

3. Don't make too many small releases. Group your releases into reasonable quantities in relation to both the manufacturer's production program and your own installation program. However, your releases should not be so large as to confront you with security and storage problems and require excessive handling.

The release form is shown in Fig. 11.2. It is a five-part set and the copies are distributed as follows:

1. Supplier's copy
2. Contract manager's copy

DRY-CORE TRANSFORMERS

Transformer no. _____

Plan nos. _____

Specifications _____

Taps _____

Connections:

   Type of lugs _____

   Size of lugs _____

Flexible connections _____

Specified sound rating _____

Testing requirements _____

Windings:

   Copper _____

   Aluminum _____

Paint finish and color _____

Shipping requirements _____

Other requirements _____

FIXTURES

Fixture type no. _____

Plan nos. _____

Specifications _____

Paint finish _____

Paint color _____

Lens type _____

Ballast type _____

Socket type _____

Voltage _____

Flexible connection:

   BX or wire and Greenfield _____

   Connection plate _____

Test data _____

Type of ceiling _____

Shipping requirements _____

FIGURE 11.1   **Purchase order check list** *(continued).*

WIRE

Type _____

Voltage _____

Jacket _____

Cable pulling eye _____

Color for phasing:

    Black _____

    White _____

    Red _____

    Blue _____

Color for grounding: Green _____

Multiple reeling _____

Reel size _____

Reel tagging _____

FIGURE 11.1   Purchase order check list *(continued).*

3. Costing copy

4. Accounting copy

5. Jobsite copy

A purchase order release is made up by the project coordinator or project manager on site and is sent to the contract manager at head office. Releases are numbered consecutively for each purchase order and are made up as follows.

Step 1. The project manager or project coordinator fills out the release order by inserting the following information: *(a)* purchase order and release number, *(b)* date of release, *(c)* name of supplier and address, *(d)* shipping address, *(e)* required delivery date, *(f)* quantity and description of material being released, and *(g)* tagging and packaging instructions.

Step 2. Copies 1 to 4 in the set are sent to the contract manager at the head office, who takes this complete set to the purchasing department for perusal and signature. The purchaser establishes the value of the release in cases in which the release does not pertain to unit-price items. Copy 1 of the release is sent out to the supplier for action.

Release Order

PO no. _____   Order no. _____

Date of release	To
Page of	
	Attention
Target	
This release	Ship to us c/o
Previous releases	
Total releases to date	Required delivery date

Item	Quantity	Description	Price	Per	Disc.	Extension

This release refers to the above purchase order number and is subject to the terms and conditions contained therein. Please make sure that the released material is delivered to the jobsite on the required delivery date and in the manner designated.

_____
Signature

FIGURE 11.2   Release order form.

Step 3. The contract manager completes copies 2, 3, and 4 by inserting the pricing information that establishes the value of the release. The clerical staff does the extensions. The release is finalized by filling in the target amount of the purchase order, the value of this release, the value of the previous releases, and the total amount of releases to date.

Step 4. The contract manager retains copy 2 and forwards copy 3 to costing and copy 4 to accounting. This procedure tells you how much of the purchase order has been released and will help accounting to process the supplier's invoices after they have been checked against their respective receiving reports. It also allows costing to determine the extent of your purchasing commitments and the balance to complete. This is required for cash-flow and job-cost analysis.

PURCHASE ORDER BULLETINS

To keep track of changes to a firm total-price purchase order, use a system of bulletins consecutively numbered. This is a four-part set:

1. Supplier's copy
2. Office copy attached to original purchase order
3. Contract manager's copy
4. Job copy

Changes to a firm-price purchase order for equipment may flow from approved change orders, job work orders, or job conditions. When this occurs, the contract manager reviews the details with the purchaser, and a bulletin is made up covering the change. This bulletin is numbered with the next consecutive number pertaining to the given purchase order, and the purchaser will mark in the price to cover the change according to the detailed description appearing in the bulletin, which he has negotiated with the supplier. After the purchaser has signed the bulletin, the first copy goes to the supplier, the second copy is attached to the original purchase order, the third copy is retained by the contract manager, and the fourth copy is sent to the job.

At suitable intervals, the purchaser accumulates all the bulletins for a given purchase order, reconciles the total change with the supplier, and issues an amended purchase order to cover these changes. In this way the purchaser will have a complete record of the changes and will be able to reconcile the final value of any given firm-price purchase order.

JOB REQUISITION

A job requisition is a notification to the purchasing department to order a given quantity of material as described therein and to have it delivered to the jobsite at any given time.

Most of the identifiable material called for in the plans and specifications, such as power and distribution equipment, raceways, wire, and cable, will be purchased in bulk by the purchasing department from data supplied by the project management. This will flow from the estimate, as reviewed and coordinated to suit the project. Job requisitions will thus cover mostly miscellaneous material, some of which may be obtained from your warehouse.

To prevent duplication of ordering and the duplicate ordering of overdue back orders requires a systematic job office routine, up-to-date filing, and adequate communication between jobsite and office. The requisitioning procedure and communications between job and office will be simplified if you adhere to the following directives:

1. Orders are not to be phoned in, unless in an emergency.

2. The requisition should be filled in neatly, giving a clear description of the items required. The requisition becomes the warehouse slip, and a copy will be returned to the job with the material. All back-order information will be listed on this return copy.

3. The material should be ordered with sufficient lead time. Plan ahead and cut down panicky, rush-delivery orders. Mark in the required delivery dates.

4. Make up your own catalog of repetitive items, such as connectors, tape, pulling compounds, fasteners, and nonelectrical items commonly used in your jobs. These items should be ordered in bulk and stocked in your warehouse. By standardizing the miscellaneous materials, whenever possible, you will cut down confusion and cost and expedite delivery of these orders, which are usually small in quantity but very frequent. By ordering from this catalog, the job and the office shipper will be looking at the same data, and this will result in a more efficient and accurate system of satisfying the job's requirements for small and miscellaneous materials.

5. Do not order similar items too frequently. Plan your work so that the number of requisitions is cut down to a controllable minimum.

6. When in doubt, check with your office contract manager before ordering material that may already have been ordered or that may be incorrect.

Although the total value of materials ordered by job requisitions is usually a fraction of the cost of the total materials required for the job, they consume a major portion of the paper work and the time spent to order, check, reconcile, follow up, pick up, and deliver. Consequently, your job planning should be directed toward minimizing, simplifying, and improving the procedures for dealing with job requisitions.

The requisition is a five-part set, illustrated in Fig. 11.3, and is filled out and distributed by the job as follows:

Job name					Job no.	JR number	Date ordered			Time ordered	Date shipped		
							Day	Mo.	Yr.		Day	Mo.	Yr.
Date required	Quantity ordered	Quantity shipped	Back order quantity	Description			Purchase order no.		Supplier		Expected delivery date		
Shipping order no.		Received above merchandise in good condition					If ship direct to job, indicate by X. ⟶						

FIGURE 11.3   Job requisition form.

Step 1. All the material which is to be requisitioned is entered in the front sheet (copy 1) in neat and legible writing or printing. This sheet and the three attached copies are snapped out as a set and sent to the contract manager at head office. Copy 5 remains on the job as their record.

Step 2. The contract manager checks the requisition to ensure that there are no errors and that it is not a duplication of material already ordered, and passes it on to the stores or warehouse. The storekeeper will endeavor to fill the requisition as much as possible from stock or from readily available sources. The materials thus assembled are noted on the requisition, and copies 1 and 4 accompany the materials to the jobsite. These sheets act as a shipping slip and show which of the requisitioned items have been shipped to the jobsite. Copy 1 is retained by the job for their records, and copy 4 is signed by the job and returned to head office as the receipt copy.

Step 3. Copy 2 goes to the costing department for pricing and filing in the cost files.

Step 4. Copy 3 is the back-order copy. When necessary, this copy goes to the purchasing department, which issues purchase orders for those items not filled by the stores. The copies of these purchase orders are attached to copy 3 (the back-order copy) and sent back to the job. This notifies the job that their requisition has been fulfilled and that the back orders are covered by purchase orders. There is a cross-reference of PO number and requisition number on the respective forms for future checking, if necessary.

In order to reduce the number of back orders and the number of purchase orders that will be required to fulfill a requisition, group similar types of items on requisitions. For example, order nonelectrical items, such as screws, fasteners, bolts, cutting oil, drill bits, tape, plywood, paint, and hacksaw blades, on a separate requisition. Don't mix them together with electrical materials.

SHORT SHIPMENTS AND BACK ORDERS

Short shipments and back orders have been interfering with the progress of your jobs for a long time. To solve a problem, you must first identify what it is. Very often material arrives in boxes and crates, and the exact quantity is not checked against the purchase order or release order until some time after the packing slip has been signed. By the time the short shipment is reported to the office, too much time may have elapsed for them to rectify the shortage with the supplier and the carrier. Many times the quantity received is marked on the receiving report, and the office only discovers the short shipment when the receiving reports are reconciled with the purchase-order quantity. This may happen months after the fact.

Many times you order standard material from a supplier, who ships part of the order to the job and back orders the balance—or the supplier may back order the complete order. The job assumes everything is under control and material will arrive as scheduled. They only become aware of the back order when they are ready to use the material and discover that it wasn't shipped. This causes delays and panics. The job and purchasing may forget about the back order and reorder the missing material, causing duplicate shipments.

How do you improve the situation? This requires initiative and a responsible attitude by all parties concerned. Every short shipment must be reported immediately to the head office. Every back order must be examined at the source, and when necessary the order should be canceled and reordered from another supplier who can better meet the job delivery requirements.

Arrange, if possible, to have a contract manager in the head office service those jobs that are not large enough to have a site staff coordinator. All information regarding short shipments or back orders is relayed by the job to the office contract manager assigned to it, whose function will be to record, follow up, and see that these matters are looked after. If you are a small contractor, you are probably carrying out the functions of a contract manager as described previously.

There are times when you are responsible for subcontract work which you prefer should be done by a specialty subcontractor. There

a) Supply (and install)_____in accordance with com-

Description of equipment and/or work
plete plans and specifications of_____ specifically sections

Name of Engineer
_____, pages _____, and addenda _____ and your quotation
_____. Guarantee period for material and labor shall begin after accept-
ance of work by the Owner and/or his representatives and shall be in effect
for a period of _____ year(s) after this date. _____ Complete bound
sets of typewritten and/or printed instructions for operating and maintaining
equipment shall be furnished, and these shall be revised as directed by the
Engineer until approval is given. This order is subject to the complete approval
of the Owner and/or his representatives. Shop drawings in_____

Number of copies
shall be submitted for approval within two weeks of the date of this order. This
equipment and/or work must comply with the latest requirements of the governing inspection authorities. Abide by the General Contractor's (Critical Path)
schedule for erection and completion of the work. (Be responsible for all cleaning, hoisting, cutting and patching related to your installation.)

Method of shipment, delivery dates (and installation dates) shall comply with
the instructions of our site personnel to be detailed later. Special notes and
consideration:

_____

_____

_____

_____

This order shall include but not be limited to the following itemized breakdown
(list may be attached to PO):

_____

_____

_____

_____

The following applicable checklist items are included in this purchase order for
materials and equipment:

_____

_____

_____

_____

FIGURE 11.4 Purchase order for subcontract work or for supply of major equipment.

are also times when you must purchase a complete system, including supply, installation, wiring, and testing. In such cases, your purchase order should be made up in the form of a subcontract as illustrated in Fig. 11.4. It is important, in these cases, to tie down your subcontractors to the general conditions, performance requirements, and completion dates called for in the governing plans and specifications.

12
Material Storage and Handling

NEED FOR PROFESSIONAL STOREKEEPERS

Handling and storage of materials on jobsites do not receive sufficient thought, planning, and in-depth management. Unfortunately, they are too often treated as low-priority, low-status operations. Both employers and the union very often think of this function as a refuge for older electricians or young helpers. Instead of realizing that professional storekeepers and specially trained personnel are required, they very often relegate this most critical and vital operation to the least productive, least trained, and least motivated. It takes people properly trained and equipped to manage and control this intricate and costly aspect of your operation. Unfortunately, many of the costs resulting from this operation are hidden costs, and what you don't see directly you do not act upon directly.

You probably expend about 20 percent of your job labor on material handling. Along with the direct labor used to unload deliveries of conduit, cable reels, and equipment, there is the loss of worker-hours when operations have to be interrupted to do this unloading, or when workers have to wait for material that has not arrived on time or for incorrect material to be exchanged, or when material that has been mislaid must be searched for.

In order to minimize this lost time and to control the amount of time used for jobsite material storage and handling, careful planning is required, starting from day one of the project. The purchaser should fully understand and be aware of the cost implications of this activity and should incorporate planned shipping and delivery requirements into the order to suppliers. This means that the project management should inform the purchaser of critical requirements, such as scheduled delivery dates, desired shipping methods (on pallets, in cartons, or strapped in bundles of sizes that are easily handled), and maximum sizes of reels and equipment sections.

A particularly serious problem is the delivery of large quantities of conduit and wire to the jobsite too early in the construction schedule. This material must often be shifted from storage place to storage place as each area becomes unavailable because of construction requirements, thus wasting many worker-hours of time and effort. Large stockpiles present security problems, such as theft, pilfering, attrition of all kinds, and possible damage. All of these are costly. Since records are very often inadequate, you most likely are not even aware of what you are losing and how much it is costing you.

PLANNING, LOCATION, AND TYPES OF STORAGE FACILITIES

Project management should therefore be very aware of the importance of proper, in-depth planning for material storage and handling. Storage facilities should be located as close as possible to where the work is being done. If it is evident that you will be forced to move your storage, use trailers which can be easily moved. If permanent areas are available, use modular sections to build the storage facilities. These modular sections can be easily assembled, and the storage facilities can be enlarged when required. They can also be easily dismantled and reused on other projects. In the interest of security, make sure that the exterior of your storage facilities is properly illuminated and also protected by a large battery-powered gong which is wired to all doors and windows —or use any other adequate burglar-alarm system. Use a flexible, modular system of shelving, which can be altered to suit the requirements of the job. Make sure that the storage is kept neat and tidy at all times.

Plan to limit on-site storage to a workable minimum, while still maintaining adequate supplies to sustain the required tempo of work. You want to avoid excessive small deliveries and pickups, and also loss of time in waiting for materials.

Examine the use on your jobsites of up-to-date secondary storage facilities, such as metal tool and material storage boxes and containers. The shipping industry has revolutionized the handling and shipping of materials with the introduction of containers, and you should look into this concept as it might beneficially suit your purpose. Tool boxes and material boxes should be of the upright, two-door type, mounted on casters and fitted with lifting hooks. They should have adjustable shelving and be secured with heavy-duty locks. Material boxes can be made larger than tool boxes and be rectangular on all sides. All boxes should bear your company name and be identified with a number.

GETTING MATERIALS TO THE WORKING CREWS

The intent is to keep the working crews working. The problem is to get the required materials to them. The old way is for the journeyman to wander over to the stores to pick up material, or else to send a helper. Often one helper is designated to pick up materials for a crew of five or six workers. This method can be easily abused, by omission or commission. One doesn't have to think too far ahead on the job if it is only a matter of sending a helper to the stores as the material requirements present themselves. This can be very inefficient and time-wasting, or it may be very pleasant to take a stroll to the stores to socialize a bit.

The delivery system is a more businesslike, efficient way of getting material to the working crews, if it is applied in a flexible manner and with judgment. The key to the success of this procedure is the crew foremen, provided they are well motivated, disciplined, and responsible. In this system, the foreman must plan and order the materials ahead of time. The storekeeper receives the order, fills it, and delivers it to the foreman's crew by using a delivery person, or "runner." This procedure can be very efficient if all concerned work to make it so. It can also break down in practice if the foreman does not plan the work and order the materials on time. The storekeeper may not deliver the materials on time, and the workers will lose time waiting for them to arrive. It takes planning, effort, and desire to make the system work. If these ingredients exist, then the delivery system is the superior system.

ROLE OF STOREKEEPER

The storekeeper ensures that the stores are clean and tidy and that materials are properly identified. Countless hours are lost searching for materials, or sorting out material mix-ups, or installing and then having to dismantle incorrect equipment that had not been properly identified.

The storekeeper ensures that shipments are checked before accepting them into the stores. If that is impossible, identify the number of boxes and mark on the packing slip, "Contents not checked." These contents should be checked as soon as possible, in order to be able to correct defects on time. The storekeeper controls the inventory and keeps track of back orders and rentals. Much unnecessary expense is incurred because rental equipment is not supervised or sent back on time.

Job storekeepers perform a function just as important as that of the foreman. They receive and correctly store material for efficient and controlled dispatching to the working crews. They expedite back orders and make sure that sufficient material is available in the stores to service the active operations. They enforce discipline around the storeroom and do not allow loitering or casual entry. They discourage the workers from coming to the warehouse for materials, except in emergencies and for other agreed-upon reasons, and they make sure that the material is delivered to the working crews on time. They follow your company procedures for receiving and returning materials and tools and keep their records up to date. All deliveries and exchanges of materials are covered in writing on appropriate forms.

You should arrange with your suppliers to deliver their material in accordance with your instructions and schedules. If deliveries arrive contrary to your instructions or at inopportune times, the carriers must be told to return at the proper time. Plan the delivery of large, heavy, and bulky loads so as to avoid last-minute, costly activities or dislocations.

The storekeeper is involved in job planning and meetings and, together with the foremen, ensures that material (particularly stock material) does not run short. The storekeeper follows up on promised deliveries and expedites back orders.

SECURITY

Arrange with the storekeeper to tour the job at the end of each day to check for materials and tools not properly secured. All infractions should be dealt with and corrected. All tools not returned to the shop at day's end should be securely locked inside the tool boxes provided for this purpose, as should materials such as small spools of wire, boxes, fittings, and devices. The larger items, such as ladders, tripods, scaffolding, and large cable reels, should be chained to an immovable object such as a column, using a heavy-duty chain and padlock.

When you have to store large quantities of conduit without proper lock-up facilities, arrange to paint the conduit ends with a chosen color for easy identification, and strap them in countable bundles large enough to discourage unauthorized persons from helping themselves to your stock. Look into the use of movable containers for conduit and wire storage.

An effective security system for your material and tools is basically a question of using common sense and caring. Take the time to plan what must be done, and follow up regularly to see that your instructions are being carried out. Every theft must be reported to the police, and to the head office in writing, giving complete, detailed descriptions and quantities. Lost, missing, or stolen tools must be reported on a missing tool report as shown in Fig. 12.1.

Make sure that tools are handed out only in exchange for a tool tag or chit, which is then placed on the tool board. Keep a tool record on file cards showing item, description, date, and name and badge number

---

Missing Tool Report

Date _____

Job name _____   Job no. _____

Tool no. _____   Serial no. _____

Manufacturer _____

Catalog no. _____

Date tool last accounted for _____

Date tool last out on site _____

Tool out to (worker's name) _____

                    Badge no. _____

Foreman's name _____

Remarks _____

_____

_____

_____

_____

_____

_____

Police officer's name and badge no. _____

Does tool require replacement?   Yes _____   No _____

Cost of loss or damage ⟶ | $ |

Note: Attach the tool slip to this report. Send the report to head office
c/o Tool Department.

                Project manager _____
                                        Signature

FIGURE 12.1   Missing tool report form.

of the worker holding the tool. When workers leave the job, they must turn in their badge, chits, tools, and hard hat to the storekeeper, who will give them a signed release for these.

ROLE AND FUNCTION OF A CENTRAL COMPANY WAREHOUSE

No doubt, it is to your advantage to cut down your central, head-office warehouse to a bare minimum, since it represents a large cost in space and personnel. However, a central warehouse will continue to remain a necessity, for some reasons which you can control and others which you cannot.

Everyday, repetitively used items which are constantly needed on your jobs should be kept in stock. You can purchase such items in bulk at the best possible prices. You can cut down the number of purchase orders and pickups for these items, and you can gain time that would otherwise be lost waiting for your suppliers to ship them. You can thus ensure that hours are not lost on the jobs when workers run short of these standard materials. Using practicality and efficiency as your guides, have as many materials as possible shipped to the job directly from your suppliers. However, you will learn from experience what standard materials you should stock in your central warehouse and in what quantities you should stock them.

Invariably at the end of a job, materials are left over for a multitude of reasons. Much of this material cannot be returned to suppliers because it may be slightly damaged, or the packages may have been removed or torn, or the items may be mixed in odd sizes and quantities. There is no alternative but to return them to your central warehouse to repair and clean them if necessary, and to list them on a proper inventory list, so that you can use them on other jobs. Make sure that your project people are aware of what is available in your stock, so that you can move this material as fast as possible.

On some jobs, storage facilities are limited or unavailable, and you will be forced by purchase agreements with your supplier to take material into your central warehouse and store it there until the job is ready for it. You may be forced to take in material to protect yourself against price increases or to cut down costly movement of this material on the jobsite. You may be confronted with serious security problems on a jobsite, particularly for expensive communication or testing equipment and spare parts, and decide to store these in your central warehouse until the job is ready for them. In other words, you should use your central warehouse in the most logical and businesslike manner to back up your jobs and protect your costs. It may pay you to have your supplier store your material, if you can arrange it, or you may have to use your own facilities. In any event, make sure that the material is properly identified and properly stored, so it will be readily available when required. Above all, plan what you do. Don't just let it happen.

Plan your storage facilities so that when materials, tools, ladders, and tool boxes come back from jobs, they are not dumped in an open yard where in a few months they will turn to junk. They should be repaired if necessary, properly stored, and placed on a current inventory list so they will be ready for use when you require them.

Deliveries from your central warehouse to your jobs should be planned and scheduled. Work out a delivery route so that your truck is used efficiently. The jobs should discipline themselves to work within this delivery schedule except for emergencies. After all, a delivery schedule is maintained in the case of out-of-town jobs.

PROBLEMS TO OVERCOME

The tendency is to deal with material handling and storage in a spontaneous manner, as it happens, rather than in a planned manner. This results in running around, searching, waiting, and last-minute ordering, all of which result in lost time. Or else there is an "I don't give a damn" attitude, or a lack of concern, which results in tools and materials being left lying around and in tool and material boxes being left open, particularly during break periods and lunch hours.

In the case of the "fetch-it-yourself" system, workers will invariably claim that they are searching for material or are on their way to the stores if they are discovered away from their work. With the runner system, workers will stand around and wait if some items of material have not been delivered on time. A well-planned and controlled material storage and handling system will eliminate these cost-consuming practices.

The project manager should definitely enforce discipline with regard to the closing of tool boxes, the security of material and tools, and the requirement for proper delivery and checking procedures. For example, foremen should be advised in a routine fashion that their orders for materials have been fulfilled and delivered. They should make out a three-copy job requisition and deposit it in an in basket at the stores. The storekeeper then fills the order. One copy of the requisition is returned to the foremen when their orders are delivered. The third copy is filed in a back-order file for action to obtain material not in the store. The back-order information is marked on the foremen's copies so that they are aware of what material is missing. Before the foremen put their requisitions into the in basket, they should stamp them with the time at which they were presented. The storekeeper should also stamp them with the time when the orders are fulfilled. This will be a record that the orders are being acted upon on time. In the case of emergencies, the foremen should write their requisitions on a form of a different color, so that they will be filled more quickly.

All problems arising from the storage and material-handling operation should be noted and reviewed, and action should be taken to solve them. By using common sense and a caring approach, you will be able to work out detailed material-handling procedures to suit your operation. As stated before, time studies indicate that about 20 percent of all worker-hours are consumed by material handling on jobs. It is extremely important to reduce this percentage, in order to expend as much time as possible on productive, direct installation activity.

Cutting down the time lost in material handling demands planned, coordinated action by both the individual contractor and the total job organization. Storage space is usually limited at jobsites, and access to it is very often difficult. This results in double and triple handling of materials as they are shifted from one spot to another. Time is wasted waiting for hoists. Truck deliveries are often unplanned, and lines and traffic jams result at the material hoists. A material-handling and traffic-control policy should be established on every jobsite to plan the storage and handling of materials, schedule hoist time, and prevent congestion and traffic problems. One of the priority activities of the project organization should be to plan the material-handling system and review it constantly to achieve maximum efficiency.

RECEIVING REPORT

Before the head office can approve a supplier's invoice and release it for payment, clearance is required from the job assuring that the material covered by the invoice has been received in accordance with the terms of the purchase order. The job fills out a receiving report, illustrated in Fig. 12.2, as follows.

Step 1. Check the quantity of pieces or cartons received, to ensure that they match the quantities listed on the packing slip or transport slip before signing for the receipt of this material.

Step 2. Report to head office immediately by phone if you discover discrepancies with the purchase order or if there are broken or missing parts. If arrangements are then made with head office or with the supplier to accept this material, note these clearly on the packing slip and in the remarks column of the receiving report.

Step 3. Check the exact quantity and the contents of the materials received, and enter these on the receiving report. Do not copy the quantities from the packing slip; you must check them for yourself. If it is impossible to check the quantities, mark the quantity of cartons which you have received on the packing slip and add the note, "Contents not checked." When you make up the covering receiving report, however, fill in the exact quantities.

Receiving Report

Date		Receiving report no.	
Job name		Page of	
Job no.		Pieces	Boxes
Supplier		Cartons	Sacks
Order no.		Bundles	Reels
Shipper		Coils	Other

Quantity	Description	Remarks

Note: List in the remarks column any material which is missing, does not conform to the packing slip, or is damaged. Fill out and send the accounting and costing copies to office immediately.

Received by _____
Signature

FIGURE 12.2   Receiving report form.

Step 4. To fill out the receiving report, write in the job name and number, the name of the supplier, the purchase order number, the name of the transport company, the date that the material was received, and the quantity of cartons, bundles, and reels, and list the exact quantities of materials which you have received. After completing the receiving report, sign it, attach the packing slip, and forward these to the head office immediately. Do not delay the transmittal of receiving reports to the head office.

Step 5. If for any reason the packing slip is missing or has been mislaid, do not hold back the transmittal of the receiving report until the packing slip has been found. Forward the receiving report to the head office immediately, with a note in the remarks column that the packing slip is missing.

Step 6. The above procedure must be followed through so as not to impose a considerable burden and loss of time on the accounting department at the head office, whose staff would otherwise have to make phone calls and do additional checking in order to clear the invoices for payment.

RETURN/TRANSFER REPORT

When materials, tools, or reels are returned to your warehouse or to a supplier, or are transferred to another job, you record this transaction on a return/transfer report. This is a four-part form, illustrated in Fig. 12.3. Copy 1 goes with the shipment, copies 2 and 3 are sent to the head office, and copy 4 remains on the job for their record. It is filled out as follows.

Step 1. Check off in the appropriate box whether the materials, tools, or reels are being returned or transferred.

Step 2. Fill in the job name and job number. Write in clearly where the materials, tools, or reels are being returned or transferred to, with the exact address.

Step 3. List the quantity and description of the items being returned or transferred. Pricing and extensions will be done at head office, and the resulting amounts will be credited to the job.

Step 4. A separate report is made up for each type of return or transfer. For example, tools should not be mixed in with reels or materials. Tools should be returned or transferred on one report, reels on another, and materials on another.

Step 5. The person on the job who is responsible for the return or transfer of the materials signs and dates the report after filling it in as detailed above. The driver or the party who picks up this material must sign for the receipt of same before it is released from the job.

EQUIPMENT RENTAL REPORT

Invariably, rented equipment is allowed to remain on the site for longer periods than required. In the meantime, the rental charges mount up.

☐ Return
☐ Transfer

Report for

Material ☐
Tools ☐
Reels ☐

No. _____
(Mark this number on credit note.)

Date _____

From (job name)	To
Job no.	
Invoice no.	
PO no.	Attention Job no.

Item	Quantity	Description	Price	Per	Disc.	Extension

Return tag no. _____

Company _____

Shipper _____
          Signature

Received by _____
          Signature

FIGURE 12.3   **Return/transfer report form.**

Field people are occupied with so many details that keeping track of rentals may become a hit-or-miss affair.

All rentals should be listed on the equipment rental report, which is illustrated in Fig. 12.4. This form should be examined on a regular basis by the project manager to ascertain which of the rented tools or equipment is no longer required and should be returned. The return of rented equipment is registered on a material/equipment return form, and a signature is obtained as a receipt for this return.

Equipment Rentals

Job name_____   Job no._____   Date_____

Item	Rented from	Date received	Period	PO no.	Date returned	Return no.

Storekeeper_____
                        Signature

FIGURE 12.4   Record of equipment rentals form.

VEHICLE DAMAGE REPORT

Every time that a vehicle is damaged you must fill out a vehicle damage report, as shown in Fig. 12.5. Write in a short but precise description of the event, stating all known details. Give the name of the worker involved and the foreman. Keep a record of these occurrences, so that over a period of time you can assess the attitudes of your workers and foremen toward company property. You will be able to identify those who are irresponsible or insensitive to this important cost item.

Vehicle Damage Report

Date _____

Job name _____ Job no. _____

Type of vehicle _____ License no. _____

Report of accident or damage _____

_____

_____

_____

_____

_____

_____

_____

_____

Name of driver	Driver's signature

Police officer's name and badge no. _____

Precinct _____

Remarks _____

Cost of damage ———————➤ $ _____

Note: Attach all relevant data and send this report to head office c/o General Superintendent.

Project manager _____
Signature

FIGURE 12.5   Vehicle damage report form.

13
Tools

In your continuous efforts to increase productivity, you rely on adequate tools to help you achieve this goal. Since only 35 to 50 percent of the working day is spent on actual direct work activity, it is vitally important to ensure that this time is utilized as efficiently and productively as possible. It would therefore be very shortsighted on your part to skimp on tools or to settle for second-rate tools. By the same token, utilizing tools that are more sophisticated and complicated than what the work requires can be counterproductive and needlessly expensive. You must use good sense and good judgment in matching the tools to the job.

You choose tools to increase the efficiency of the work, to reduce the amount of time taken to complete an operation, and to ensure good-quality workmanship. The proper tools will reduce fatigue and thus increase production. To maximize the amount of time in the day for the direct work activity requires that you decrease the amount of time spent on material handling. The investment in tools and equipment that allows you to achieve this will be rapidly repaid if you choose them wisely and in sufficient quantity.

ACCOUNTING FOR THE COST OF TOOLS

An analysis follows of the types of tools used, and their cost, on a large, fast-track project for the government which was completed by Electrical Contractor E at the end of 1978.

The scope of work included power distribution, lighting, electric heating, related controls, and motor connections. The final value of the contract was $3.3 million. The duration of the contract was 24 months,

TABLE 13.1   DISTRIBUTION OF WORKER-HOURS ON A LARGE PROJECT

WORK ACTIVITY	TOTAL WORKER-HOURS
Conduit work and cable trays	56,000
Wiring and cable installation	27,000
Fixtures	10,000
Balance	24,000
Supervision	30,000

and 147,000 worker-hours were consumed in the work. These worker-hours divided up as shown in Table 13.1. The following tools and equipment were used to perform the work.

Equipment and Tools for Job Office and Workers' Shacks

1	60-ft trailer
2	Punch clocks
5	Filing cabinets
3	Drawing racks
36	Drawing sticks (clamped holders)
5	Office desks
9	Office chairs
2	Drawing tables
1	Typewriter
1	Photocopy machine
1	Calculator
1	Label-making machine
2	Water coolers
5	Office stools
8	Wardrobe cabinets
8	Tables
16	Benches
2	Refrigerators

5   Hotplates
1   Air conditioner
60   Hard hats
1   20-in electric fan
2   5-kW heaters
6   Fire extinguishers
2   Movable workers' shacks

The total monthly rental value of all the above equipment utilized in the job office and workers' shacks worked out to $650.

Equipment and Tools for Material Handling

1   60-ft trailer
46   Tool and material storage boxes
36   Conduit buggies
10   Material storage baskets on wheels
6   Farm wagons
1   Forklift truck
1   Battery-operated electric cart
2   Hydraulic pallet lifters
1   Compact, lightweight hoist
4   Reel jacks
4   Crowbars
600 ft Chain

The total monthly rental value of all the above equipment and tools utilized for material handling worked out to $1300.

Lifts, Scaffolds, and Ladders

1   Scissor lift to 42 ft
4   Scissor lifts to 25 ft
1   Mini scissor lift
8   Electric toppers (lift to 34 ft)
4   Manual hydraulic toppers (lift to 24 ft)

40   Scaffold sections, complete
13   Mobile adjustable-height rolling miniscaffolds, complete
18   6-ft ladders
10   8-ft ladders
 4   12-ft ladders
 2   30-ft aluminum ladders
 2   20-ft catwalks

Much of the work had to be done at above normal heights. The total monthly rental value of all the above equipment worked out to $3800.

Tools for Conduit and Raceway Installation

These were also used for other operations.

13   Mechanical benders, up to 1-in rigid conduit
 2   Mechanical benders for 1¼- and 1½-in rigid conduit
 5   Power drives with stands
 2   Portable power drives
 2   ¾-in electric drills
10   ⅜-in electric drills
20   ½-in hammer drills
 2   ½-in hickeys for rigid conduit
23   ¾-in hickeys for rigid conduit
16   1-in hickeys for rigid conduit
 1   1¼-in hickey for rigid conduit
17   ¾-in EMT benders
 6   1-in EMT benders
 6   1¼-in EMT benders
 9   Portable band saws
 1   Saber saw
14   Chain tripods
 3   Screw guns
15   ½- to 2-in hydraulic knockout sets
 3   ½- to 1-in hydraulic knockout sets
 2   2½ to 4-in hydraulic knockout sets

10	½- to 1¼-in hand knockout sets
1	Bolt cutter
14	Pipe cutters
18	Pipe reamers
7	Ratchet handles with ½- to 1-in dies
19	Ratchet handles with ½- to 2-in dies
13	24-in levels
1	Welding machine
10	½-in box wrench sets
14	24-in pipe wrenches
6	36-in pipe wrenches
4	36-in chain tongs

The total monthly rental value of all the above tools, which were required for the installation of conduit, tray, and other operations, worked out to $1500.

Tools for Wire and Cable Installation

1	Hydraulic cable bender
1	Hydraulic cable cutter
2	Cable pulling machines
2	Come-a-Longs (Chicago grips) with 100 ft of steel rope
4	24-in pulleys
5	18-in sheaves
4	No. 652 tray sheaves
26	No. 658 tray sheaves
5	Pulling-wire baskets
1	Manual lug squeezer
20	Wire skinners
7	Reel stands
4	Wire racks
4	Reel jacks

The total monthly rental value of all the above tools, which were required for the pulling in of the wire and cable, worked out to $400.

Miscellaneous Tools and Tools for Testing

1	Industrial vacuum cleaner
45	Safety belts
20	Light stands
5	Hand lanterns
4	Walkie-talkies
2	5000-V meggers
1	Amprobe tester
1	600-V megger

The total monthly rental value of all the tools and equipment required for miscellaneous activities and for testing worked out to $350.

The grand total monthly rental value for all tools and equipment, under all headings above, works out to $8000. Contractor E expended $150,000 for tools and equipment on the above project. If you work this out as a percentage of the contract price and the total labor cost and also as a cost per worker-hour you arrive at the following:

Expenditure for Tools

- Five percent of total contract price, or
- Ten percent of total labor cost, or
- $1.00 per worker-hour

The above analysis covers tools for material handling, for installation and connection of the materials that go into the project, for testing, and for job housekeeping. Contractor E also maintains a well-equipped prefabrication shop. The tools employed for shop work are covered in Chap. 10.

The point of the exercise is that you should work out what the costs are for tools, so that you can bill for them in your cost-plus work and allow for them in your estimates. The above figures can be used as a guide. Keep in mind that expendable tools, such as hacksaw blades and drill bits, are not included in your capital cost for tools but are treated as a material cost.

All tools wear out. They must be repaired and maintained, and eventually they must be replaced. At the price that you are paying for a worker-hour, it is illogical to work with unsatisfactory or decrepit tools. The time wasted fussing with these tools and sending them back to the shop, and the resulting disruption of the work, will cost you many times more than what you would have paid for satisfactory tools in the first place.

You should designate one worker to be in charge of tools. Although you have to educate all your workers to respect and look after the tools, nonetheless there has to be a person in charge who will enforce the rules, to ensure that the jobs have the right tools and that they are looked after in the right way.

MAJOR PROBLEMS ASSOCIATED WITH TOOLS

Problems associated with tools are proper use, maintenance, loss, and safety. Don't take for granted that your workers automatically know how to use the tools properly. The proper skills should be taught, as to someone learning to drive a car should be taught. Arrange for frequent demonstrations and classes for your foremen and key personnel, where they will receive instructions from experts.

Overtooling, like overeating, becomes counterproductive and wasteful. Some operations may require sophisticated tools like hydraulic scissor lifts. When absolutely necessary, they can be invaluable in getting the work done efficiently. However, it is a costly decision to use such equipment when ordinary scaffolding or ladders will do. Because of strict safety regulations, hydraulic lifts cannot be moved when a worker is on the work platform. The worker must descend from the platform, the platform must often be lowered to clear an obstruction, the lift has to be moved to the new location, and the lift must then be raised again. This sequence of activities is time-consuming, particularly if it has to be repeated again and again. Also, the more sophisticated the equipment, the more often it may tend to break down, resulting in lost time while workers stand around waiting for the necessary repairs to be made. Your workers will begin to demand the sophisticated equipment as necessities, when in fact simpler tools may be much more economical and practical. Judgment and control must therefore be exercised in the selection of the right tools and scaffolding for a particular job. As in most judgments and decisions, the simplest choices, carefully thought out, are usually the best ones.

Tool breakage and maintenance costs are very high for most contractors. Much of this stems from a poor attitude, or lack of precise knowledge in the use and care of tools, on the part of many workers. Control of this cost is a management responsibility; without it you will suffer serious losses. The person in charge of tools should not release a tool until it has been cleaned and repaired. Repair includes replacing broken or missing parts, sharpening cutting edges when necessary, and checking electrical parts for grounds or faults. As you well know, tools are becoming more sophisticated and more expensive, and it is in your interest to educate your personnel in their proper use, care, and control.

Most tools are lost through negligence and lack of concern. Every tool loss should be reported on a missing tool form, as illustrated in Fig. 12.1. This report lists the details of the loss, the worker and the foreman involved, and the value of the tools. In this way you will have a record of the loss frequency for your various workers and foremen. If patterns of negligence or dishonesty appear, the culprit should be warned, and if the situation is not rectified, the person involved should be dismissed. The procedure of filling out the form when a tool is lost requires that a thorough search and check should be made to establish who last used or saw the tool. Was it loaned out? You should find out who was working close by, the method of storing and handling the tool, and whether in fact there was a break-in. In such an event, notify the police.

A record card in duplicate is made up for every tool. This card lists the complete history of the tool, with these details:

- Name of the tool and general description, including size, model type, and serial number
- Name of the manufacturer and the supplier, including address, phone number, and date of purchase
- Record of maintenance and repairs, including dates, and costs incurred
- Your own shop number assigned to it, which should also be stamped on the tool itself

One of the cards remains in a master tool file, and the other card goes into the job tool file when the tool is sent to a job.

A shipment of tools from the shop to a job is covered by a three-part tool slip. Copy 1 is kept in the shop. Copies 2 and 3 accompany the tools to the job. Copy 2 is signed for by the job and is returned to the shop, where it is attached to the original in the job file. Copy 3 remains on the job for their records.

The return of tools and their transfer from job to job are covered in Chap. 12. On medium and large jobs, the stores usually assign a secure area for tools. There are a master tool list and a tool board to record and keep track of the tools. Tools are only released in exchange for a tool chit which is placed on the tool board. The person to whom the tool is released and the date of release are recorded. This procedure is detailed in Chap. 16.

Many accidents occur because tools are improperly used or maintained and because of a sloppy attitude on the part of the people involved. For example, construction workers have insufficient knowledge about the correct type of safety belts to be worn and the correct manner of wearing them and attaching the lanyards. Materials such as leather and manila rope, which have been discarded by safety committees, are still in use.

Tests have shown that, when a fall is arrested by a safety belt fastened to a lanyard exceeding 4 ft in length, the person falling may sustain serious injury. If a lanyard is fastened to the side of the safety belt, whip action can result. When fastened to both sides of the belt, it can cause back injury. The safest way of attaching a lanyard is at the back, since during a fall the body will bend forward naturally. It is the responsibility of the employee and the foreman to inspect the safety belt carefully and to refuse to wear it if it is flawed, frayed, damaged, or otherwise not up to acceptable standards.

The lesson of this example applies to the choice and use of all tools. The proper attitude about the selection and use of tools not only will save worker-hours, but also may save lives.

TOOLING UP

To run an efficient and productive job you must have the right tools and the right attitude about their use and care. Here are some rules for effective tool management on a project.

1. The tool manager at the head office has a master card system that lists each tool owned by the company, including its statistics. All tools are numbered.

2. A tool shipped to a jobsite from the shop is accompanied by two parts of a three-part tool slip. One is signed and returned to the tool manager, who attaches it to the original and places it in the job file. The other is kept by the job as a record of tools on site.

3. All requests for tools should come from the project manager or foreman.

4. Tools should be ordered well in advance, so as not to cause delay if the tools in question are not in stock and must be obtained from other projects or purchased.

5. Project managers planning to increase their work force should review the project tool inventory to ensure that there will be sufficient tools on site to meet the requirements of a larger crew. Similarly, when a crew is decreased, surplus tools should be sent back to the shop.

6. The foreman should check the condition of the tools in use and make sure that they are maintained and cleaned periodically to prevent breakdown and delay. Broken tools must be returned immediately to the shop for repairs and replacement.

7. Whenever possible, each team should have a storage box for tools and material in the area where they are working.

8. The foreman should ensure that each team is knowledgeable in the use of tools issued to them.

9. The electrician, not the helper, is responsible for the proper use, care, and storage of any tool issued to him or her.

10. Tools must not be transferred from one job to another without the authorization of the tool manager. A transfer will not be authorized unless the tools are in proper working condition.

11. The project manager should maintain and enforce a tool tag system to keep track of tools.

12. The loss of a tool must be reported immediately to the tool manager, who should investigate the matter, ascertain responsibility, and institute corrective action when necessary.

The following standard tools are used on all jobs and are usually available as required:

- Metal tool and material storage boxes on wheels
- Conduit wagons
- Working benches on wheels
- Dollies, 30 × 30 in, good for up to 3000 lb
- Step ladders of all sizes, also on wheels
- Scaffolding of adjustable height and width, with standard wheels and with car-type trailers
- Pipe benches and vises, bench vises
- Ratchet dies, ½ to 2 in
- Dies 2½ in and up, machine-driven if permission is obtained from head office
- Power drives for threading (require head-office permission)
- Hickeys, all sizes, rigid and thin-wall
- Standard drills, ¼- to ¾-in chuck
- Redhead drill hammers for ¼- to ½-in shields
- Electric screw guns up to ½ in

- Electric impact box wrench
- Pipe cutters, plumber's hand style
- Cutting and welding equipment, gas
- Cup saws, all sizes
- EMT offset benders, ½ and ¾ in
- Pipe reamers, all sizes
- Hydraulic knockout punches, ½ to 2 in, 2½ to 4 in
- Pipe wrenches, 1 in and over
- Cable cutters, hand, hydraulic to 1000 MCM (1 million circular mils)
- Bolt cutters, ¼- to ¾-in bolts
- Compact, lightweight hoists, 1½ to 5 tons
- Power-actuated tools (if allowed), or piston-driven tools
- Jet-line vacuum fishing equipment
- Welders, electrical, 220 and 550 V
- Hand winches for cable pulling, 2 to 5 tons
- Snatch blocks, various sizes
- Amprobe testers, various current ratings
- 600- and 1000-V meggers
- Telephone handsets, for testing
- Battery lanterns
- Metal cabinets for up to sixty-six 1000-ft reels of no. 12 wire (These are locked up in the cabinet when the day's wire pulling is done.)
- Metal baskets on rollers for material handling

The following special tools are available for use on the projects upon request, subject to approval at the head office:

- Scissor lifts, up to 25 and 42 ft
- Mini scissor lifts and manual hydraulic lifts, up to 17 ft
- Cantilevered lift platforms (electric toppers), up to 34 ft
- Electric cable pullers
- Sheaves for cable pulling
- Reel rollers, skid type
- Reel jacks, up to 10 tons
- Track jacks, toe jacks, 5 to 10 tons

- Press drills for bench mounting
- Skill saws, bench saber-tooth saws, portable band saws
- Sheet-metal shears.
- Diamond and core drills, vacuum and fixed-weight, for drilling concrete
- Hydraulic benders, for up to 5-in conduit, subject to approval from head office.
- Angle- and flat-iron hydraulic or hand punches and cutters
- Hossfield hand benders, column-mounting type
- High press lug and connector, hydraulic or hand tools
- One-shot offset or standard benders, electric-driven, ½ to 1¼ in, subject to special approval by head office
- Sheet-metal nibbler
- Underfloor duct insert locators
- Extension ladders up to 40 ft
- Special ladder fixture-lifting assembly with counter balance
- Roller-type dollies, 8 × 8 × 4 in, 7½ ton capacity, with steering attachment
- Pothead kits
- High-voltage wire insulators, gloves, boots, mats, etc.
- 15-kV hook stick tester
- 5000-V megger
- Avro meters, for small current and resistance testing
- Electric winches, various speeds and capacities, for cable pulling or hoisting
- Material hoisting rigs and assemblies
- Any other special tools, not listed above, subject to head-office approval

14
A New Look at Temporary Wiring

THE NEED FOR MODULAR, REUSABLE COMPONENTS

Temporary wiring for construction projects is an ever-increasing expense item. Is there a way to control this cost? The answer is yes—if you start treating temporary wiring as you treat tools, some of which are consumed on the job, but most of which are designed to be reused and transferred to other jobs. This can be achieved by utilizing standard, modular, movable units, ruggedly constructed, that can be easily joined together in smaller or larger temporary wiring networks as the job requires.

Temporary power and lighting are necessities that you hate spending good money for. As jobs get bigger and safety codes become more demanding, the cost for these services keeps growing and growing. For years, money has been poured down the drain for this operation, to pay for materials and worker-hours that are rarely salvaged. Thousands, and sometimes hundreds of thousands, of dollars are spent for temporary power and lighting of a construction project. What is salvaged at the end is mostly junk, or else it becomes junk after the temporary wiring materials have been dumped in your yard. On your next job, when you again have to install temporary wiring, you can never find what you require in your stock.

When you begin to look at temporary wiring in the same way as you look at tools, as items which are required on every job to enable the project to be built efficiently, you will see it in an entirely new way. The way is to standardize the components and put emphasis on reusability and flexibility to suit varying job requirements. When you are through with the temporary system, you can send these units back to your shop where they will be cleaned up and repaired, if necessary, and stored for use on your next job. The money spent for constructing the modular units will thus be recovered from the many jobs on which you will use them, and in the long run will greatly reduce the cost of the temporary wiring.

There are three basic types of units that you require for a reusable temporary wiring system:

1. Power unit assembly
2. Satellite plug-in unit assembly
3. Temporary lighting stand

POWER UNIT ASSEMBLY

The components of this assembly are shown in Fig. 14.1. A 100-A 480-V service entrance switch feeds a splitter trough with one 100-A and one 60-A disconnect switches. The 60-A switch feeds a 45-kVA dry-core

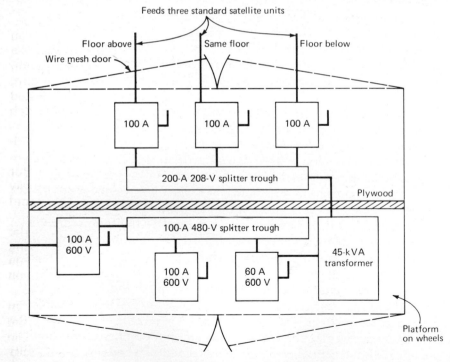

FIGURE 14.1   **Layout of power unit for temporary wiring.**

transformer, leaving the 100-A switch available to service other 480-V temporary requirements. Or the 100-A 480-V switch can feed a 75-kVA dry-core transformer, which will then leave the 60-A switch available for other temporary requirements.

The secondary of the 45-kVA transformer feeds a 200-A 208-V splitter trough with three 100-A disconnect switches. Each of these disconnect switches can feed three satellite plug-in unit assemblies, which are described further on. In a typical temporary wiring arrangement for a highrise building, one of these disconnects feeds three satellite plug-in assemblies on the same floor as the power unit, one disconnect feeds three satellite units on the floor below, and the remaining disconnect feeds three satellite units on the floor above. A 45-kVA power unit can thus service three floors of temporary wiring, as shown in Fig. 14.2. When a 75-kVA transformer is used, the secondary splitter box is 400 A and feeds five 100-A switches, as illustrated in Fig. 14.4.

The power unit consists of a ventilated cabinet 60 in wide, 30 in deep, and 60 in high. The 60 × 30 in base rests on four sturdy wheels. The front and the rear of the cabinet are accessible by wire mesh doors. The top of the assembly is a solid metal pitched roof which overhangs the cabinet and contains two heavy lifting eye brackets, one on each side. The base of the cabinet is bisected by a ¾-in plywood partition that runs the length and height of the cabinet. The 480-V switches are installed on one side of this plywood partition, as illustrated in Fig. 14.3, and are accessible by one set of doors. The 208- or 220-V switches are installed on the other side of the plywood partition, as illustrated in Fig. 14.4, and are accessible by the other set of doors. The dry-core transformer rests on the floor of the cabinet.

SATELLITE PLUG-IN UNIT ASSEMBLY

This unit consists of a plywood backplate which is fastened to a tripod. The tripod can be collapsed for shipping and storage. A 24-circuit lighting panel is installed on the plywood surface. Eight duplex plug receptacles, four on each side, are installed in the side walls of the lighting panel, and each is fed from a separate circuit. Two 2-pole 220-V 30-A circuit breakers are installed underneath the lighting panel and connected to it. These breakers are located external to the lighting panel in order to obviate having to open the panel to connect a 208-V load when required. Over the top of the lighting panel is a box for connecting the feeder from the power unit and the feedthrough to the next satellite panel. There is also a box for connecting the strings of temporary lighting pigtails or temporary lighting stands. The satellite plug-in unit assembly is illustrated in Fig. 14.5.

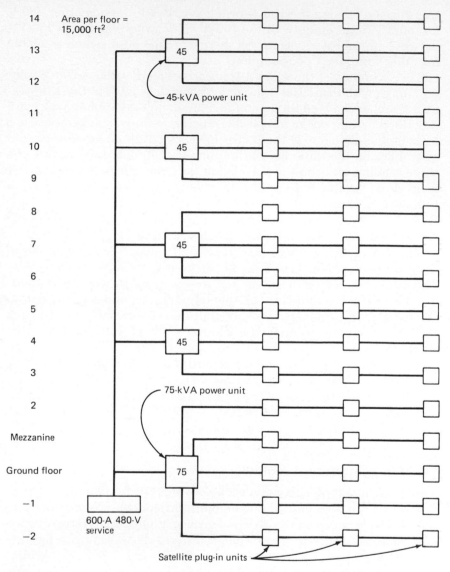

FIGURE 14.2   Typical temporary wiring network for an office building project.

TEMPORARY LIGHTING STAND

The temporary lighting stand is adjustable from 8 to 12 ft. This is accomplished by arranging for the top 1-in rigid pipe to slide inside the bottom 1¼-in pipe. The bottom pipe is supported by a heavy metal base plate. The top pipe is slotted and slides up and down along a pin which

FIGURE 14.3   Power unit for temporary wiring (front view).

FIGURE 14.4   Power unit for temporary wiring (rear view).

FIGURE 14.5 Satellite plug-in unit.

is welded on the inside of the 1¼-in pipe and prevents the 1-in pipe from being removed. A heavy chain is welded to the base plate so that the light stand can be secured to the nearest column. Two 500-W quartz lamp fixtures are mounted at the top of the light stand, as shown in Fig. 14.6. Wire guards should be installed around the fixtures to prevent the lamps from being stolen.

TYPICAL TEMPORARY WIRING REQUIREMENTS

The building blocks of the reusable temporary power system are utilized as shown in Table 14.1.

A string of pigtail sockets spaced 20 ft apart with a 150-W bare lamp in every socket is the most common method of supplying temporary lighting on a construction site. However, there are many problems connected with the use of this method, which requires excessive maintenance. The strings of pigtails get in the way of construction and are easily damaged. In the case of highrise buildings, the strings are in the

**TABLE 14.1 COMPONENTS OF A
REUSABLE TEMPORARY POWER SYSTEM**

ITEM	AREA COVERED, FT^2
45-kVA power unit	45,000–90,000
75-kVA power unit	75,000–150,000
Satellite plug-in unit	5,000–10,000
150-W lamp	400
Light stand with two 500-W lamps	2,500

way of the ceiling installers and very often get pushed into the false
ceilings. When the structural steel is sprayed with fireproofing, the
pigtails and lamps get coated with this material. Lamps are constantly
broken or stolen. A large amount of labor is spent to repair, relamp, and

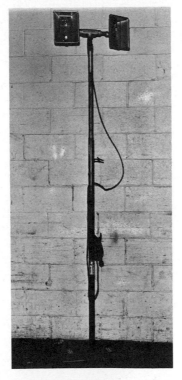

**FIGURE 14.6 Light stand for
temporary lighting.**

maintain these strings of temporary lighting pigtails. Light stands are more flexible. They can be easily moved to where they will not interfere with the work or where they are most required. Above all, there is a saving in labor because relamping and maintenance are greatly reduced.

In general, the requirements for temporary wiring on a jobsite are as follows. One pigtail socket with a 150-W lamp is needed for every 400 ft² of floor area. Usually the pigtails are spaced at 20-ft intervals along prefabricated strings 200 ft in length and are suspended 6 in below the ceiling. Additional pigtails (about 20 percent more) are usually required because of areas that are enclosed during construction. A 150-W lamp is provided over every landing in the stairwells. In the case of machine rooms or higher-bay areas, a 200-W lamp is provided for every 300 ft² and is suspended about 12 ft above the floor. A 150-W lamp is required at every elevator landing, and a 200-W lamp is installed in the elevator shafts at every second floor. A 1000-W floodlight is installed 15 ft above street level over loading platforms.

A lighting panel is usually required for every 5000 to 10,000 ft² to feed the temporary lights in that area, and also for an assembly of receptacles for small electrically operated tools. Usually, two 30-A 220-V switches are called for in this assembly, for tools that may require this voltage.

Temporary wiring connections are required for hoists and for traveling cranes. Sometimes, the permanent elevators have to be connected temporarily, either for testing or because they can be used for hoisting construction material and crews. Some projects call for temporary heating with gas- or oil-fired unit heaters. These may require temporary wiring connections for the blower motors of the unit heaters.

You may also be confronted with substantial costs to bring in a service of adequate capacity to feed the temporary power requirements of the construction project. Large projects may require a temporary high-voltage substation. All of the items which may be required for an adequate temporary wiring system add up to a substantial cost. In addition to the cost of the installation itself, you are also faced with the cost of continuous maintenance of the system. Not only are worker-hours consumed in normal maintenance activities, but many are lost due to disruption of the work when the temporary lighting fails or when plugs are found to be inoperative. When the temporary wiring is part of your contractual obligation, you are often forced to disrupt operations and to spend worker-hours answering service calls to replace burnt-out lamps, repair broken wires, and troubleshoot the many real or imagined problems that develop.

It is therefore evident that you should spend time to analyze your experiences with temporary wiring and to come up with a system of rugged, modular, reusable assemblies that will reduce maintenance and allow you to recover the cost of the equipment over many jobs. You can figure that you will expend about 5 worker-hours per 1000 ft^2 of project for temporary wiring. To estimate the cost of temporary wiring, you can use the following formula as a rule of thumb:

$$\text{Cost of temporary wiring} = \frac{\text{area in (ft}^2)}{1000} \times 5 \text{ worker-hours}$$
$$\times \text{ worker-hours multiplier} \times 2$$

For example, if the total area of the project is 500,000 ft^2 and your cost per worker-hour (worker-hour multiplier) is $20, then the estimated cost of the temporary wiring will be

$$\frac{500,000 \text{ ft}^2}{1000} \times 5 \text{ worker-hours} \times \$20 \times 2 = \$100,000$$

You can use this formula for budgeting purposes or to arrive at a ball-park figure of the cost of temporary wiring.

TEMPORARY USE OF PERMANENT WIRING

Very often in the later stages of a project, the general contractor or the owner may wish to use the permanent electrical system for temporary power. If this request is made on your project, you should obtain a signed agreement beforehand that protects you. This agreement should cover the following points:

1. The building should be closed in and made watertight before your switchgear, panels, and fixtures are energized.

2. You should receive a preliminary acceptance of your installation by the engineers, before it can be utilized for temporary power.

3. All warranties on the equipment and installation should commence when they are put into use.

4. You should be paid for all subsequent cleaning and repairs that may be required for the equipment used for this purpose.

5. You should be paid to replace the lamps used for temporary lighting, if required, including the labor expended to replace them. Otherwise, you should be given a written confirmation that the lamps in use will be accepted as is.

6. You should not be responsible for theft, abuse, or damage to your equipment used for temporary purposes.

7. You should not be charged for any consequential losses due to the disruption or failure of the power used for temporary purposes.

8. All maintenance and repairs to the permanent equipment and wiring resulting from their use for temporary power should be done by your forces.

9. You should be compensated for any subsequent repairs and shutdowns which may be required in order for you to complete your work in accordance with the plans and specifications, or for any overtime work that may be required to achieve this.

Before you allow your permanent installation to be used for temporary wiring purposes, make sure that you obtain agreement from an authorized representative on as many of the above points—and as many other points as are pertinent and applicable.

15
Job Office Procedures

On small projects, the job office procedures and paper work are usually divided between the job foreman and head office. Larger jobs usually require a full-time clerk. In all cases, the job office procedures and paper work should be as simple as possible and should be standardized so that your involved personnel will become expert in carrying them out.

Office procedures involve keeping records and making out reports. Information coming into the office or transmitted out of the office must be filed for easy access and retrieval. A planned program covering records, reports, and filing will contribute to efficient job management. Organized procedures to achieve this should start from day one of the project. Otherwise, the paper will pile up, and you will find it difficult and time-consuming to locate the desired information quickly and easily when you require it.

FUNCTIONS OF A JOB CLERK

On a large job, the following functions are performed by a job clerk. On a small job, these functions, or whatever portion of them is required, are divided between the foreman and the contract manager at head office.

1. Reports and records relating to hours worked.
   a. Make up the time sheets.
   b. Look after the punch-in and punch-out cards.
   c. Deal with payroll data, information problems, and errors.
   d. Keep track of apprentice books.

    e. Keep records of employees who arrive late, leave early, or are missing.

    f. Fill out forms for new employees and, when required, arrange for photos and passes.

    g. Prepare tool tags for the employees.

    h. Make up accident reports.

2. Transmittal and filing of drawings.

    a. Record all incoming drawings and ensure that the drawing record is kept up to date.

    b. Make up transmittal sheets for all drawings sent to the general contractor, construction manager, engineers, and suppliers.

    c. Make up covering transmittal sheets for all samples sent to general contractor, construction manager, and engineers.

    d. File all drawings and shop drawings.

3. Work order and change orders.

    a. Record, file, assemble, and transmit to the head office all work orders and covering vouchers. Make sure that all work orders and vouchers are properly signed.

    b. Record and file all change-order quotations. Keep track of submissions and approvals and inform the project manager when a change is approved.

4. Reports and records.

    a. Make up reports for transmittal to the head office, as described below.

    b. Keep records of truck use by noting drivers' names, mileage, gas consumption, and repairs.

    c. Make up and transmit all reports required by the general contractor or construction manager (that is, relating to areas worked in and associated personnel).

    d. Keep track of rentals.

    e. Keep track of expense vouchers.

    f. Make up and distribute minutes of staff meetings.

5. Filing.

    a. File correspondence, memos, minutes, and reports.

    b. Keep the delay file up to date.

6. Miscellaneous.

    a. Answer the phone and keep records of important calls.

b. Make copies of documents and transmit to the people who must receive them.

c. If there is no full-time storekeeper, or if the storekeeper is too busy, the clerk may be involved in chasing up back orders, sending out the prefabrication sheets, checking that the requisitioned prefabrication material is delivered on time, and filing purchase orders.

REPORTS AND RECORDS OF HOURS WORKED

The recording of time worked is a basic job office function. Larger jobs are usually furnished with a time clock. This clock should be installed in a convenient location where the office clerk can exercise surveillance over punching in and punching out. The time cards are sent to head office at the end of each week. On smaller jobs, the daily hours worked are recorded by the foreman.

Weekly Labor Distribution Sheet

On both large and small jobs, the hours worked during the week are entered on a weekly labor distribution sheet, which is illustrated in Fig. 15.1. This sheet is sent to the payroll department to make up the weekly pay for the workers. It must therefore be an accurate report of the exact number of hours spent by each worker on the project. It is filled out as follows:

1. Write in the job name, the job number, and the date on which the week ends.
2. Write in neatly and clearly the name and badge number of each worker on your project.
3. Write in the number of regular hours worked by each employee for each day of the week.
4. If it is necessary to work overtime, fill out a separate labor distribution sheet for this purpose.
5. The sheets must be signed by the project manager or job foreman before being sent to the office.
6. All payroll calculations will be done at head office.

Weekly Labor Report

The project is broken down into measurable operations which are listed on the weekly labor report. This report is illustrated in Fig. 15.2. For each working day of the week, the total hours expended on each active

Weekly Labor Distribution

Job name _____ Job no. _____ Week ending _____

Name	Qualifications	Badge no.	Hours worked							Total hours	Regular hours	Overtime hours	Rate	Travel		Gross amount
			Mon.	Tues.	Wed.	Thurs.	Fri.	Sat.	Sun.							
Totals																

Foreman _____ Signature

Project manager _____ Signature

FIGURE 15.1 Weekly labor distribution sheet.

		Weekly Labor Breakdown Report								

Job name_____   Job no._____

Foreman_____   Week ending_____

Code no.	Description	Daily tally							Total worker-hours this week
		S	M	T	W	T	F	S	

FIGURE 15.2   Weekly labor breakdown report form.

operation on which work has been done during the day are entered into the daily column adjacent to the given operation. The total hours on this report must equal the total hours contained in the weekly labor distribution sheet from which the weekly pays are made up.

The weekly labor report is completed by writing in the job name, the job number, and the date on which the week ends. It is signed by the project manager or the job foreman and sent in weekly to the personnel manager at head office. A copy is retained for job records.

Summary Labor Report

The list of operations shown on the weekly labor report is repeated on the summary labor report. This form is illustrated in Fig. 15.3.

The estimated worker-hours for each operation are listed in the initial target column. At the end of each month, the total hours consumed by every active operation are totaled up from the weekly labor reports and entered on the summary labor report under "total worker-hours used to date." Each operation in this column is a running total which increases every month by the number of hours consumed by each active operation during the month.

The labor hours included in the approved changes are apportioned to the various operations, and the revised hours are entered in the corrected target column. In the percentage completed column, you enter the total percentage up to which each active operation has been completed at the end of the month. By multiplying the hours in the corrected target column by the total percentage completed, you obtain

Summary Labor Report

Job name_____   Job no._____

Foreman_____   Month of_____

Code no.	Description	Initial target	Extras	Corrected target	% complete	Worker-hours earned	Total worker-hours used to date	Worker-hours gained	Worker-hours lost
Total									

FIGURE 15.3 **Summary labor report form.**

the worker-hours earned for each operation. By comparing the hours earned with the hours used, you establish the worker-hours gained or lost.

The summary labor report is completed by entering the job name, the job number, and the date on which the month ends. The form is signed and sent in monthly to the personnel manager at the head office. A copy is kept for job records. This report will give the project manager and the head office an indication of how the work is progressing. It can be a forewarning of a deteriorating situation so that corrective action can be taken in time.

It is important that the worker-hours be listed against the actual operations on which they are spent. You should instruct your field people not to juggle the percentages and figures in order to make the report look good. This report is like a thermometer. Its purpose is to inform you if the patient is healthy or running a fever. By keeping it accurate, you will help to identify problems in sufficient time to effect a possible cure.

Job Percentage Completion Report

The job percentage completion report is used for progress billing. It is completed by the twentieth of each month and sent to the contract manager at head office. The form is illustrated in Fig. 15.4.

Job Percentage Completion Report					

Page_____ of _____

Job name _____ Job no. _____ Date _____

Item code no.	Description	Cost breakdown by item or code	% done		Amount
			This period	To date	

Signature

FIGURE 15.4   Job percentage completion report form.

The contract is broken down into logical categories which generally follow the scope of the work. The amount of each category includes both material and labor. The total of all the categories adds up to the total contract amount. The categories and the respective amounts are listed in the job percentage completion report. The approved changes are listed at the end of the form for billing purposes.

Around the twentieth of every month, the project manager or job foreman fills in the percentage completed against every active category and writes in the job name, the job number, and the date on which the month ends. You should see to it that materials and equipment delivered to the job during the month are placed in their proper location and that their physical presence on the job is reflected in the percentage completion. You can obtain the value of this equipment from the contract manager at the head office to help you derive the required percentage completion.

This report is the basis of your monthly progress billing. When you compare the monthly billing derived from this report with the actual cost appearing in the costing records up to the end of the given month, you will have a check on the performance of the job.

FILING AND RECORDS

The following systems of files are kept at the jobsite office.

1. General file. Separate folders properly labeled contain the following:
   a. Correspondence. On large projects, it may be necessary to use separate folders for correspondence to and from others. All correspondence must be stamped with the date on which it is received.
   b. Transmittal sheets. On large projects, it may be necessary to use separate folders for transmittals to others and from others.
   c. Weekly project reports.
   d. Job percentage completion reports.
   e. Billing records.
   f. Minutes of job staff meetings.
   g. Record of rentals.
2. Supplier file. This file contains copies of all shop drawings and catalog cuts for which approval is required. A separate folder properly labeled is kept for each type of equipment and material category.
3. Change-order file. A separate folder is made up for the following:
   a. Record of change orders.
   b. Record of work orders.
   c. Change-order quotations, in numerical order.
   d. Work-order vouchers.
4. Material file. This file consists of binding cases or folders labeled as follows:
   a. Purchase orders completed.
   b. Purchase orders outstanding.
   c. Requisitions completed.
   d. Requisitions outstanding.
   e. Back orders to be followed up.
   f. Material release forms.
   g. Receiving reports.
   h. Wire and cable purchase orders.

    i. Major equipment purchase orders.

    j. Special material purchase orders.

    k. Balance of purchase orders.

    l. Material and tool transfer slips.

    m. Tool slips and tool reports.

    n. Reel record.

5. Labor file. This file consists of binding cases labeled as follows:

    a. Weekly labor distribution sheets.

    b. Weekly and monthly labor reports.

    c. Productivity reports.

    d. Accident reports.

    e. Minutes of safety meetings.

6. Claim file. This file consists of binding cases with the following contents:

    a. Forms for reporting partial or total stoppage of the work.

    b. Copies of correspondence dealing with delays and factors having an impact on the job that may potentially cause you damage.

    c. Copies of memos and instructions that have an impact on the scheduling and performance of your contract.

    d. Copies of weekly project reports and minutes of job meetings dealing with delays and impact factors.

    e. Copies of productivity records and reports relating reduced productivity to job impact factors.

In addition to maintaining files in the six categories listed, your job office keeps track of drawings, transmittal sheets, change-order quotations, work orders, and changes to equipment. Descriptions follow of procedures for keeping these important records.

Drawing Record

To ensure that your installers are working with the latest drawings at all times, arrange to maintain a continuously updated drawing record. At the beginning of the job, list the original tender drawings, including their dates and revision numbers. As changes are made and revised drawings are issued, record these revisions and list the approved dates of the respective change-order quotations to establish that these latest revised drawings are approved for construction. Make sure that your field people are aware of all changes, the status of approvals, and the latest drawings with which they should be working. Figure 15.5 is the drawing list form.

Drawing List

Job name_____ Job no._____

⊠ Contract   C− Change notice   ◤ Approved   ◤ Not approved

Drawing	No.	Revision
		0	1	2	3	4	5	6	7	8	9	10

FIGURE 15.5   Drawing list form.

Transmittal Sheet Record

All submissions of shop drawings, samples, and reports are accompanied by a transmittal sheet. The basic function of this sheet is to give you a comprehensive record and a receipt confirming that the noted items were in fact transmitted. The transmittal sheet is a record of the drawings or catalogs sent to engineers, inspection authorities, suppliers, general contractors, and construction managers for the purpose, as noted therein. This three-part form, illustrated in Fig. 15.6, is filled out as follows:

1. Write in the name and address of the party to whom the drawings or data are being transmitted.
2. Check off the purpose of the transmittal in the appropriate box.
3. Write in the date, the job name and number, and the number of copies to be returned.
4. List the drawing numbers, including their respective titles, dates, and issue and revision numbers. Write in how many copies of each drawing are being submitted.

Transmittal Sheet

To		
Attention	Return	copies

Submitted for the purpose noted

Inspection approval ☐	Refused ☐	For information ☐
Engineer approval ☐	Returned with corrections ☐	For quotation ☐
For approval ☐	For construction ☐	As-built ☐
Approved with corrections ☐	Additional copies ☐	_____ ☐

Job name	Job no.	Date

Item/document no.	Revision	Date	Copies	Description
Remarks				

Date received _____   Received by _____

Signature

FIGURE 15.6   Transmittal sheet.

5. Write in any pertinent remarks, if applicable, and sign the transmittal form.

The transmittal form is distributed as follows: Copies 1 and 2 accompany the transmittal. Copy 1 is signed by the receiver and returned to the head office for their records. Copy 2 remains with the receiver of the transmittal. Copy 3 is retained on the job and filed in the appropriate transmittal sheet file.

Change-Order Quotation Record

The head office maintains a record of all change-order quotations. Each submission is given a change-order number in consecutive order. The job staff maintain a duplicate record for their own information, so that they can follow up and expedite approvals and can ensure that the work is being done in accordance with the latest approved changes. This record tabulates the change-order number, the date submitted, the amount of the submission, the date approved, the amount approved, and the owner's approval number.

Work-Order Record

The same procedure is followed for work orders as for change orders. The record tabulates the work-order number, the date billed, the amount billed out, the amount of payment received, and the date on which payment was received.

Changes to Equipment Record

The estimating department at the head office keeps track of the changes brought about by change orders to the originally tendered equipment. This usually involves equipment such as fixtures, panels, and bus duct; systems such as fire alarm, communications, and energy control; and motor controls and switchgear. The estimating department keeps an updated record for each type of equipment, which lists the types and quantities tendered and all changes in types and quantities brought about by change orders. This gives a running tally which can be reconciled for the purposes of amending the purchase order and ensuring correct material releases from the job. Regular updated copies of these records are transmitted to the job for their information and are filed in the material files.

ADDITIONAL WORK AND CHANGE ORDERS

The Work Order

A work order is a contract authorizing you to do a specific amount of work, usually on a time-and-material basis. Before you do this work, make sure that you obtain an authorized signature on the work order. Verbal orders are not acceptable.

The work order is a four-part form, illustrated in Fig. 15.7, which is made up as follows:

1. Enter all the pertinent information on copy 1 in neat and legible writing.

   a. Write in the job name and address, job number, and date.

   b. Apply the next consecutive work-order (WO) number to this work

---

Work Order

Job name	Job no.
Job address	Work order no.
	Invoice no.
	Date

Bill to

Address

Description of work

~~~~~~~~~~~~~~~~~~~~~~~~~~~~~~~~~~~~

General conditions:
1. The party signing this work order and the associated vouchers must have full authority to do so on behalf of the company for which the work is done.
2. That party's signature on the vouchers is our confirmation that the quantities have been checked and approved.
3. Terms of payment: net 30 days.

Work authorized by (company's name)
Representative
Title
Signature

FIGURE 15.7   Work order form.

order. Only one work-order number is used for a given work order, and this number will be inscribed on all vouchers pertaining to this work order.

c. Write in the name and address of the party to whom the work will be billed. Make sure that this information is specific and correct.

d. Describe the work which you have been authorized to perform in as clear and detailed a manner as possible.

e. Enter the name and title of the party authorizing this work order and the company represented, and have this representative sign the work order.

The first step of filling in the work order is now completed. Peel off copies 1 and 2. Copy 1 is sent to the contract manager at head office, and copy 2 is given to the party authorizing the work order.

2. You now have copies 3 and 4 left in your set.

a. As the work is done and entered onto vouchers, mark these voucher numbers in the space provided for them on copy 3.

b. When the work order is completed, enter the completion date on copy 3 and sign it. Tear off copy 3 and send it to the contract manager at head office. Copy 4 remains in your work-order book as your record. Make sure that all the voucher numbers pertaining to the completed work order are entered on copy 3 before you return it to the office. It is essential that the information which you enter on the work order should be specific, precise, and accurate and that the party who signs for it has the authority to do so.

The work-order form can also be used when you are authorized to do a specific amount of work at a firm price. In that case, you enter the agreed-upon firm price in the description of the work.

The Voucher

A signed voucher is a receipt for the material and labor used in the execution of a work order. The voucher is a four-part set, illustrated in Fig. 15.8, which is filled out as follows:

1. Write in the job name and number.

2. Write in the work-order number which appears on the work order for which this voucher is made up. All vouchers pertaining to the same work order carry the same work-order number.

3. Where there is more than one voucher pertaining to a given work order, the vouchers are numbered consecutively in the space designated "sheet no." For example, if there are six vouchers pertaining to

Job name		Job no.	Date			
Bill to			Work order no.			
			Sheet no. of			
Description of work						

Quantity	Material		Price	Unit	Amount

Miscellaneous					
Truck deliveries					
Subtotal					
Tax paid at source					
		Subtotal material			
Permit and inspection					
Total unit price					
		Subtotal expense			

Classification	Quantity	Regular hours	Time and a half	Double time	Total hours	Rate	Amount
Electrician							
Helper							
Foreman							
Quantities checked and approved for payment by					Subtotal labor		

Customer _____   Foreman _____
                    Signature   Signature

FIGURE 15.8   Voucher form.

a given work order, number them 1 to 6 in the order in which they are made up. This numbering procedure, with the printed voucher numbers, is a double check that all vouchers are accounted for.

4. Under "Bill to," write in the name of the party who will pay for this work.

5. Write in a short description of the work covered by the voucher in the space marked for it.

6. Date each voucher on the actual day that the work was done.

7. List all the materials which have been used in the work covered by the voucher. Make sure that you do not forget to list the accessory and miscellaneous items that are frequently forgotten, such as tape, connectors, clips, fasteners, drill bits, saw blades, and fuses. Conduit should be charged out in 10-ft lengths. For example, if 8 ft of conduit is used, charge for 10 ft.

8. Do not price the materials. This will be done in the office. When unit prices apply, group those materials covered by the same unit price together and write in the applicable unit prices. Do not mix regular material and unit-price items. Keep them separate.

9. Under "miscellaneous," make sure to charge for

   a. Any applicable permit and inspection fees.

   b. Testing, if required.

   c. Truck deliveries. Most materials arrive on the jobsite by truck. A proportionate amount of the cost of trucking the material to the jobsite should be charged to the work order as it applies.

   d. Any other applicable expense items, such as crane rental, traveling expenses, and room and board, as they apply.

10. List the number of teams, or of electricians and helpers, and the actual hours that they worked at straight time and overtime. There are blank spaces available for other personnel who may be used in the work, such as foreman, coordinator, and clerk. The application of the wage rates and conversion into dollars will be done in the office.

11. When the voucher is completed, it is signed by your foreman and by the person authorized by the party who asked for the work to be done. All vouchers must be signed before they are sent to the office.

12. The completed and signed voucher is distributed as follows:

    a. Copies 1 and 2 are sent to the office.

    b. Copy 3 is given to the party who authorized and signed for the work.

c. Copy 4 remains on the job as their record.

13. The voucher sets come in books and are numbered. Every number must be accounted for to ensure that vouchers are not mislaid or lost. The voucher sets are numbered consecutively. Use consecutive numbers of vouchers for the same work order, as a double check that all vouchers for a given work order are accounted for.

Change-Order Quotations

Changes to the scope of work covered by your contract may come by way of addenda or other written instructions incorporated in change-order requests. Very often these changes are based on revised drawings and specifications. You must insist that all change-order requests be in writing and clearly define the changes called for.

As soon as you receive a change-order request, you must give it a change-order (CO) number. Change-order numbers are assigned in consecutive order, one per change-order request, as they arrive. Consecutive numbering is a control to ensure that all changes are accounted for. A takeoff is made for each change order and is listed on an estimate sheet, illustrated in Fig. 15.9.

The procedure for filling in these sheets is as follows. In the spaces provided, write in:

1. The job name and number, and your change-order number.
2. The name of the person who requested the change, and the date of the request.
3. The customer's reference number and date.
4. The numbers of all drawings on which your takeoff is based, including all revision numbers and their dates.
5. The list of all materials in the material column, with adequate descriptions and the respective quantities and prices.
6. Worker-hours, in the column designated for labor. Many contractors refuse to divulge their labor units and instead apply worker-hours to groupings of material, or they may use composite units relating to assemblies or operations. It is recommended that an agreed-upon method of arriving at the labor costs should be settled with the approval authority, in order to expedite the approval of the change-order quotations. The labor estimate must reflect job conditions and the actual cost of doing the work, and not some abstract system that may in fact shortchange you.
7. Estimate sheet number. All estimate sheets must be numbered consecutively.

Change requested by	Date	Job name
Customer reference no.	Date	Job no.
Drawing no.	Date	Our quotation no.
Drawing no.	Date	Date
Revision no.		Page of

Description of work

Material	Quantity	Material price	Per	Extension	Labor

Estimated by _____
                    Signature

FIGURE 15.9   Change-order estimate sheet.

When all the estimate sheets are completed, the totals are summarized on a summary sheet. This summary sheet is illustrated in Fig. 15.10 and is filled out as follows:

1. Write in the headings as directed by instructions 1 to 4, above, for filling in the estimate sheet.
2. Write in a short description of the work covered.
3. Write in the material costs of the estimate sheets and add them up. Apply the applicable sales tax percentage to this total figure.
4. Total up the worker-hours of all the sheets and enter this figure in the total worker-hour column. Multiply this figure by your labor multiplier to obtain a subtotal in dollars. Multiply this subtotal by the percentage covering your labor burden to obtain the estimated labor cost.
5. Write the direct job expenses which are applicable to this change in the job expense column.

Change Order Summary Sheet

Job name _____ Job no. _____ Date _____

Change requested by _____ Quotation no. CO _____

Customer reference _____ Sheet _____ of _____

Material								
Sheet no.								
Subtotal material								
% Sales tax								
Total material								

Labor								
Total worker-hours								
Worker-hour multiplier								
Subtotal								
% Labor burden								
Total labor								

Job expense								
% Job expense								
Bonding								
Room and board								
Total job expense								

Description and notes

Total prime cost								
% General overhead								
Subtotal								
% Profit								
Subtotal								
Subcontract work								
Unit price work								
Total price								

This quotation is subject to the following conditions:

1. This quotation is valid for____days.
2. Labor is based on regular working hours.
3. Written confirmation required before above work will proceed.

FIGURE 15.10 Change-order summary sheet.

6. Material costs, labor costs, and job expenses are added together and entered as the total prime cost.

7. To this subtotal apply your percentage to cover general overhead.

8. To this add your percentage to cover your profit.

9. Add the cost of subcontract work, if any.

10. Add the total of unit-price work, if any.

11. This will add up to the total price of the change-order quotation.

12. Make sure that all necessary explanatory notes are entered under "description and notes" for such work as cutting and patching by others or any special work that is excluded.

13. If there is a possibility that this change will have an impact or ripple effect on your main contract add the following note: *"Any costs or expenses* which may result from the impact of this change on our contract are not included in this change-order quotation and may be covered in a future claim when the facts are known."

In most cases, the takeoff, pricing, and submission of the change-order quotations are done at the head office and the job office receives copies of these submissions for their records. The change-order quotations are recorded, as previously described in this chapter. After they are approved, the approval dates are recorded and the change is incorporated into the overall work program.

DRAWINGS AND MANUALS

Coordination Drawings

The purposes of a coordination drawing or sketch are

1. To minimize interference between the trades

2. To minimize trial and error and lost time in the field

3. To achieve the simplest and most economical installation that will satisfy the purpose and intent of the design

4. To supply the installation personnel with the necessary details and information, so that the work will be installed in an efficient and neat manner

5. To incorporate the contractor's years of field experience to achieve the best possible installation at the lowest possible cost

6. To develop assemblies, particularly in the case of repetitive operations, that can be prefabricated and mass-produced

The coordination drawings should be as simple as possible, concentrating on center-line dimensions and incorporating as much additional information as is necessary for the field personnel to do their work efficiently.

There is no virtue in going the long way round or in spending unnecessary material or labor to achieve the purpose and intent of the design. You must insist on your right to apply your years of field experience to achieve the intent of the design in the most economical and efficient manner.

Runs of heavy conduit risers, bus ducts, and trays should be checked for interference and for the shortest routes, and also to minimize the number of required offsets, elbows, bends, and crossovers. Branch wiring should be rationalized and made modular, where possible, for prefabrication and mass production. Critical areas such as boiler rooms, mechanical rooms, and electrical rooms should be carefully studied to achieve the best possible arrangement of equipment and feeders, to ensure accessibility, to shorten connecting runs, and to minimize crossovers and interference.

Coordination drawings affecting other trades should be reviewed with responsible representatives of these trades, who should be asked to sign them to record their approval and agreement. You must be careful not to minimize or eliminate the role of the general foreman and the field foremen in relation to the coordination function. Many types of field coordination problems can best be handled by these experienced personnel.

The coordination team consists of the project manager, the project coordinator, the general foreman, and the area foreman. The division of responsibility should be clearly spelled out. The project coordinator should be in constant touch with the field supervisory personnel, using memos, sketches, and drawings. In turn, these field people should make up sketches or mark up drawings for your records, and should communicate with their project coordinator so that there is an efficient flow of information back and forth.

There is a lot of potential initiative, imagination, and ideas in the field, which your coordinator should tap and make use of. Basically, though, coordination requires organized reviewing and digesting of many pertinent sources of information in order to come up with the right answers and layouts to help the field people do their work in an efficient and economical manner. On small jobs, the coordinating function is carried out by the job foreman and by the contract manager at the head office who is looking after this project. On large jobs, there is usually a project coordinator assigned to this task. The qualifications of a project coordinator include a knowledge of electrical drafting and design, field experience, and an imaginative, questioning mind capable of handling and digesting the pertinent data.

Shop Drawings

Job specifications very often require the submission of shop drawings for the engineer's approval before fabrication or installation is permitted. These are required by the contractor, as well, for coordination purposes in order to preview and solve future installation problems. The three most common types of shop drawings are the following:

1. *Catalog cuts.* Items such as fixtures, starters, and devices are very often described in detail in manufacturers' catalogs. These descriptions may suffice for approval purposes.

2. *Outline drawings.* These give a general outline description of the fixtures or equipment and show pertinent dimensions and details. They may be sufficient for approval purposes if the application is straightforward. In most cases they are used for preliminary approval, subject to the approval of the final manufacturer's construction drawings.

3. *Manufacturer's fabrication drawings.* These are the final shop drawings which show in detail how the manufacturer proposes to build the equipment. All the pertinent details that fully describe the equipment which will eventually arrive at the jobsite are shown. These drawings are for final approval to manufacture.

The job keeps a record of all shop drawings, showing when they were requested, received, and submitted for approval, the date of final approval, and the date when an approved drawing was returned to the supplier. All shop drawings must be stamped with the date on which they were received. Submission of these drawings for approval or their return to the supplier must be recorded on a transmittal sheet. The job keeps a record of these transmittal sheets.

The project manager and the coordinator should carefully examine and study the shop drawings before submitting them to the engineers for approval. The details should be checked against plans, specifications, the purchase order, and job requirements. Make sure that these drawings contain all the pertinent details, such as the size of sections for shipping, location and size of openings, special fasteners or devices to lift or connect, method of shipping (such as by pallet or crate), flexible leads, lug size, type of trim, gauge of metal, identification requirements, and color. Check to ensure that all the requirements of the specifications, the electrical inspectors, and the job are met.

The cycle between first receipt of a shop drawing and its final approval can be very time-consuming, particularly if that drawing has to go back and forth several times for correction. In order to save time, the project coordinator should first check very carefully and correct the errors, ratify these with the suppliers, and submit the corrected drawings for approval. Ask for the drawings to be stamped "Approved" or "Approved as noted" by the engineers, not "Approved for general arrangement." You must insist that when corrections are of a minor nature, the engineers should stamp them "Approved as noted" to expedite the approval procedure.

When it is possible or practical, the jobsite coordinator or another designated person should visit the manufacturer during the time when your equipment is being manufactured, in order to ensure that the equipment will arrive exactly in accordance with your requirements. Much labor is lost in the field correcting manufacturing errors.

Make sure that you keep sufficient sets of shop drawings for manuals when they are required.

As-Built Drawings

Specifications require that a contractor must supply as-built information on a set of white prints which are provided for this purpose. Usually this task is left until the end of the job, when much time and effort must be expended to mark up the set of white prints with the as-built information, in order to get final acceptance and release of holdback.

The project management must discipline their foremen to mark in the as-built information on their portion of the white prints progressively and continuously, as the work is done. Then, when the project is completed, the drawings will all have been marked up. Although this is the logical procedure, it very rarely is followed, and you must insist that it be followed.

Your responsibility to produce as-built drawings does not mean that you have to bring the engineers' drawings up to date by incorporating the changes that were called for by them in sketches, memos, or field instructions. It is the engineers' responsibility to update the drawings and incorporate the changes made by them.

In order to protect yourself against inadvertent loss or destruction of the as-built drawings, it is a good idea to have them photographed on miniature slides which can be easily stored and are available to be enlarged in the event of a loss. This precaution should be applied to other valuable data that are vulnerable and would be very expensive to reproduce.

Instruction Manuals

Many projects require that you turn over, at the termination of the job, a set or multiple sets of manuals containing shop drawings and operation and maintenance information relating to the major equipment and systems supplied. Too often the project management waits until the end of the job before it starts to accumulate all the necessary shop drawings, catalog cuts, and spare-part and operation-maintenance information to fulfill this obligation. Tardy submission of these manuals can hold up final acceptance and release of holdback.

Do not wait until the end of the project to fulfill this obligation. Make it a continuous and progressive operation throughout the duration of the job, so that at the end the manuals will be complete and ready to be turned over. Use a standard hardback binder for this purpose. Make sure you request and obtain the correct quantity and type of information required. The information should be subdivided into logical sections, such as high voltage, motor control centers, panels, and fixtures, separated by proper dividers and tabulated. Shop drawings should be folded and inserted in heavy manila envelopes which have been punched to fit the binder. Check the specifications for the following requirements:

1. Quantity of manuals required
2. Whether a special format is called for
3. Requested titles for the various sections
4. Special data required, such as photometric data, lighting curves, fuse or breaker coordination curves, and test data

When you turn over the manuals, make sure they are recorded on a transmittal sheet. This transmittal sheet must be signed for by an authorized receiver of these manuals. This signed transmittal sheet should be properly filed as your record of having fulfilled your obligation with regard to instruction manuals.

Marking and Identification

Large sums of money are often wasted on a project because inadequate attention is paid to the requirements for identification and marking.

Furthermore, final acceptance is often delayed and holdback is not released until this neglect has been rectified at the end of the job.

Lack of tagging, marking, and identification of circuits results in lost time during jointing, circuit connections, and testing. The time to identify the circuits, the wiring, and the equipment is when the work is being done. At that time the circuits should be marked on a temporary card fastened to the panel. Color coding of wire should be used. The covers of boxes should be painted in different colors to identify the voltage of the system.

It is important that your project management is committed to this concept and takes the necessary steps to enforce it. Circuits should be identified in a logical and precise manner by referring to column numbers, room numbers, quadrants, motor numbers, or any other designations that will specifically tie down the identification. General references, such as "receptacles" or "lights," are not sufficient or acceptable. Read your specifications carefully with regard to identification, particularly when they refer to different sizes and colors of lamicoid plates, special panel and switchgear identification, mimic bussing, or any other special tagging requirements. Your purchase orders should incorporate and clearly stipulate these requirements when you want the manufacturer of the equipment to supply these markings and identification.

Even if the specifications do not bind you to a given system of identification, your own need to reduce the loss of worker-hours that results from inadequate tagging of the wiring, circuiting, or equipment should ensure that you cooperate with the field to achieve this requirement. This is just another part of the productive style of work that you are aiming for.

16
Procedures for Managing a Cost-Plus Project

INTRODUCTION

Most contractors, at one time or another, are awarded contracts on a *cost-plus* basis. This means that payment will be made on the basis of either cost plus a fixed percentage or cost plus a fixed fee. In all cases, you will be asked to submit a detailed procedure for recording, checking, and controlling the cost of the materials and labor that will be expended in the work. This chapter describes procedures and paper work that can be used to manage and control a large cost-plus project. It can serve as a guide for any size or type of project.

A large cost-plus project requires a self-contained organization with all the management procedures and controls that are required to run an independent business. In analyzing such a project, therefore, you have to deal with and identify many of the basic concepts and procedures that, to a lesser or greater degree, are common to all contracts, big or small.

The managerial chart shown in Fig. 16.1 is for a very large project. On smaller contracts, many of the functions indicated in this chart are combined. Thus, for example, the functions of coordinator and general foreman may be carried out by the project manager on some contracts. The functions shown in the chart must be carried out, to a lesser or greater degree, on all contracts. The quantity of personnel required to carry them out effectively depends on the size of the project. Managing the project, or managing a business, involves four basic control functions:

1. Estimating, planning, coordinating, and scheduling
2. Purchasing, warehousing, and delivering the materials to the working crews

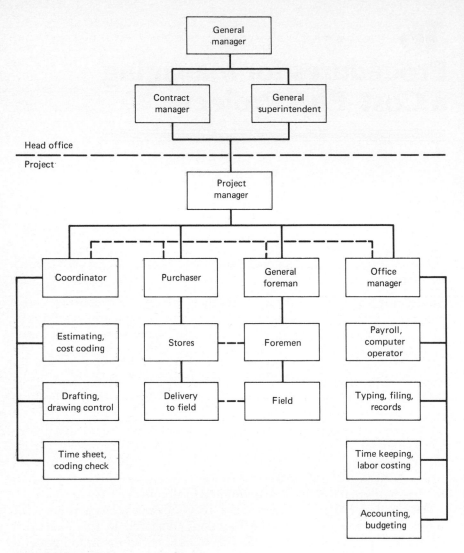

FIGURE 16.1   Management chart.

3. General and field supervision

4. Payroll, accounts payable, typing, filing, and accounting

Items 1 and 3 have been covered in previous chapters. This chapter deals with procedures and paper work to manage items 2 and 4.

TABLE 16.1 AUTHORIZED SIGNATURES NEEDED FOR APPROVALS

ACTIVITY	AUTHORIZED SIGNATURE
Field requisitions for materials or tools from stores	Foreman
Request to purchase materials	Coordinator and storekeeper
Purchase orders to suppliers	Purchasing agent and project manager
Tool purchases	Purchasing agent and project manager
Clearing incoming invoices for payment	Accounting and project manager
Payroll checks	Project manager and designated head office person
Accounts payable checks	Project manager and owner
Petty cash	Project manager

APPROVALS

The signatures of designated authorized personnel are required in order to process requisitions, purchase orders, payroll checks, accounts payable checks, petty cash disbursements, and other such activities. The authorized signatures needed in each case are listed in Table 16.1.

PURCHASING

A request to purchase materials is originated either by the coordinator or by the stores. However, the actual purchase order (PO) itself is made up by the purchaser, who arranges for it to be typed by a person designated for this task. The typed PO is submitted to the project manager for review and signature, who in turn sends it to the owner for approval and signature. The owner sends out the original copy of the PO to the supplier in the accompanying addressed envelope, retains one copy for record purposes, and returns the balance to the purchaser, who distributes the copies as detailed further on.

The coordinator originates a request to purchase by referring to the list of estimated materials, always taking into account the coordination and delivery requirements of the work program. The stores originate a request to purchase mainly for standard accessory and shelf-type materials. Such items may have been requisitioned by a field foreman and found to be out of stock, or else material that is normally stocked in the warehouse has fallen to a minimum level calling for replenishment.

Whatever the point of origin, the request is made on a four-part purchase order memo (POM) form which is signed by the originator and sent to the purchaser. The POM form instructs the purchaser to place an order for the materials as detailed therein. It therefore contains in rough draft form all the necessary information required to make up the PO, such as quantities, description, coding, and tagging and shipping instructions. The purchaser completes this draft copy by writing in the PO number, the name of the supplier, unit prices, applicable discounts, taxes, and the total price. All the POMs are numbered. The purchaser circulates this form as follows:

Copy 1	Retained by the purchaser and used by the secretary to type out the covering PO
Copy 2	Sent to stores for their records
Copy 3	Sent to the field for their records
Copy 4	Sent to the coordinator

The PO form is a seven-part form. After being typed, it is returned to the purchaser for proofreading and checking. The purchaser signs the PO and sends it to the project manager for a final check and signature. It is then forwarded to the owner for approval, and is finally distributed as follows:

Copy 1	Sent to the supplier
Copy 2	Filed by the purchaser after approval by owner
Copy 3	Retained by the owner's construction manager
Copy 4	Sent to stores
Copy 5	Sent to the cost file
Copy 6	Sent to accounting
Copy 7	Sent to the numerical file

A PO which is for less than $1000 may be printed by hand. Otherwise, all POs are typed. When ordering materials, it is important not to mix together different code materials or materials from different suppliers. Also, all pricing information, delivery dates, colors, finishes, dimensions, locations, tagging, and any other pertinent data must be clearly spelled out. For items that require quotations prior to the placing of a PO, three independent quotes are obtained and are submitted to the owner for record and approval along with the covering PO.

RECEIVING MATERIALS ON THE JOBSITE

The suppliers ship the materials to the jobsite along with a covering packing slip. The job stores receive the materials, check the items against the respective packing slip and PO, and make up a receiving report (RR). The owner's checker examines the shipment and signs the RR. One copy of the RR is retained by the owner, one copy is filed at the stores, and the remaining copy is stapled to the packing slip and sent to the purchaser for a visual check. The purchaser then passes it on to accounting.

All materials and equipment are received on the jobsite only by designated and authorized stores personnel. Every RR is given a number and logged in numerical order in a material received book, by listing the name of the vendor, the RR number, and a brief description of the material received. A new page is started each day and dated. All materials must be accompanied by a packing slip. Each RR indicates the date that the shipment was received, the packing slip number, the quantities and description of the materials received, and the receiver's signature.

JOBSITE STORES AND WAREHOUSING

Materials in the stores are properly sorted, and placed in bins or on shelves in a neat and organized manner. All bins are clearly identified, marked, and coded. The coding of materials follows a coding list which identifies the item associated with each code number. The stock in the jobsite stores is controlled, so that fast-moving items such as locknuts, bushings, clips, fasteners, boxes, connectors, and small wire are available in adequate quantity at all times. An inventory of the stock is taken periodically, and a report is submitted to the project manager for review. Materials and tools are issued only by authorized stores personnel to field personnel presenting a signed, authorized stores requisition. Materials which are too large to be stored in the warehouse are located adjacent to it in a secure manner.

The following procedure is followed for receiving materials on the site:

1. The material is checked by the warehouse receiver and by the owner's checker, who initials the packing slip.
2. If the shipping slips do not refer to the PO number, the location, etc., the stores personnel will check with the respective PO and mark the cartons accordingly.
3. The RR is filled out and attached to the packing slip. The RR is distributed as described in the preceding section.

4. When the material on a given PO is completely received, it is recorded on the copy of the PO sent to the stores, which is filed alphabetically by supplier, and on a photocopy of the PO, which is filed in numerical order. The completed PO is moved into the completed section of the supplier file. Once a week or sooner, if necessary, the stores provide the purchasing department with a list of the purchase orders which have not been completed. Purchasing compares these with their records and expedites as required, or seeks clarification for the delay.

5. Each foreman makes up a material requisition (MR) form for the items required by the crew, stipulating the appropriate code numbers, and submits it in duplicate to the stores. The MR are numbered.

6. The storekeeper makes up the order as per the MR, and records the MR number in a book. The items not in stock (back orders) are indicated on the MR and listed in the book. The order is delivered by the stores' runner to the area requested on the MR. The foreman must sign the MR to acknowledge receipt of the material.

7. A copy of the MR, which has been coded and identifies the area to which the material has been shipped, is sent to purchasing for their record and filing.

All items which are shipped directly to an area without first passing through stores are checked on arrival by the storekeeper and the owner's checker, and are subject to the same paper work as previously described.

RETURNING MATERIALS

A material return/transfer (MR/T) form is utilized to record incorrect material that is returned to the suppliers for repairs or replacement. The MR/T form lists the following information and is signed by the suppliers when they pick up the returned material:

1. PO number
2. Reason for the return and the date
3. Tender call number, when applicable
4. A complete description of the returned material
5. Original RR number

The MR/T form is countersigned by the owner's representative, who retains a copy at the time of signing. Copies of the signed MR/T form are distributed as follows:

Copy 1   Returned with material to vendor

Copy 2   Kept by owner's checker

Copy 3   Sent to purchasing for checking and to be filed with PO in the open file

Copy 4   Retained by stores

INVENTORY AND CONTROL OF TOOLS

When it is necessary to purchase a tool, the request for this purchase is made to the coordinator. The coordinator reviews the request and takes it up with higher authorities in the case of an expensive tool or piece of equipment. If the request is found to be justified, the coordinator initiates a PO for it.

As soon as the tool is received by the stores, the following information is recorded on a card and also in a tool record book:

1. Your company tool number

2. PO number

3. Purchase price

4. Date of purchase and date of receipt

5. Name of manufacturer and catalog number

6. Name of supplier

7. Description of the tool

8. Serial number of the tool

A copy of the completed record card is forwarded to the purchasing department for their file and use. The tool record book is kept by the stores. Each tool is entered on a separate page.

All requests to the stores for tools are made by the foreman only. When a tool is issued, the date of issue, the requisition number, the foreman's name, and the name of the worker to whom the tool is issued are all recorded in the tool record book on the page provided for that tool. The status of each given tool is also visually recorded on a tool-check control board, which has four check columns:

1. Out on site

2. In stock

3. Out for repair

4. Not accounted for

When a tool is received for the first time (at time of purchase), a tag is made up indicating the tool number, and the tag is then placed on the appropriate peg on the control board.

A check is carried out by the toolkeeper at least once a month on the status of all tools, and a report is submitted to the project manager with the following information:

1. Tools issued to each foreman
2. Tools in stores
3. Tools out for repair
4. Tools not accounted for

Every lost tool must be reported immediately to the project manager on a lost-tool report form. When a tool is sent out for repair, the date on which it was sent out, the name of the repairer to whom it was sent, and the nature of the repair are all recorded in the tool record book.

PROCESSING INVOICES

The supplier makes out an invoice in four parts on which are listed the PO number, the code number, the quantities and types of materials, price per unit, applicable discounts, total price, and sales tax. The invoice is sent to the job accounting department, where it is processed as follows:

1. Every incoming invoice is stamped with the date.
2. The invoice is matched against the respective PO and RR. The invoice number and the associated data are listed in the accounting records.
3. All pricing and extensions are checked. The invoice is stamped and all the required data filled in.
4. The invoice is matched with the PO, the RR, and the packing slip, and a four-part supplier voucher is made up to cover this package.
5. A check with three voucher copies is made up to cover the approved invoice amount.
6. The complete package (invoice, PO, RR, packing slip, supplier voucher, check, and check voucher) is sent to the project manager for initialing and for signing the check. The package is then sent to the owner. When only a partial payment is involved, two photocopies of the PO are sent with the invoice. The original PO is sent only with a total or final payment.
7. If an invoice is found to be incorrect in part or in total, a debit memo is made up by the accounting department to correct the error. A

copy of the debit memo is attached to the invoice when it is passed for payment. The original of the debit memo is sent to the supplier.

8. When a partial or total payment is made for a given PO, the respective RR number is noted against each item of the PO.

9. All invoices are payable on a given date of every month, except for invoices which are subject to cash discounts. These are processed immediately.

10. The accounting department records the check numbers in numerical order in the cash disbursement ledger, along with the name of the supplier and the amount.

11. The owner receives the invoice package, examines it, and signs and mails out the check together with copy 2 of the supplier voucher in the addressed envelope that accompanied the package. Copy 1 of the supplier voucher and copy 1 of the check voucher are retained by the owner. The balance of the package is returned to the job accounting department, which distributes it as follows. Copy 2 of the check voucher is filed numerically. Copy 3 of the check voucher, along with copy 3 of the supplier voucher, copy 6 of the PO, the RR, and the packing slip, is placed in a dead accounting file in chronological order. Copy 4 of the supplier voucher is filed in an alphabetical file.

12. All checks issued in any given month are entered on a monthly check-listing sheet. Checks are listed numerically and by supplier's name and amount.

CODING

The costs of all materials are identified and coded as follows:

Conduit (see Fig. 17.1)	Codes 100 to 199
Wire and cable (see Fig. 17.2)	Codes 200 to 299
Raceways and nonmetallic piping (see Fig. 17.3)	Codes 300 to 399
Accessories and finishing materials (see Fig. 17.4)	Codes 400 to 499
Fixtures and lamps (see Fig. 17.5)	Codes 500 to 599
Power and distribution equipment (see Fig. 17.6)	Codes 600 to 699

| Systems equipment (see Fig. 17.7) | Codes 700 to 799 |
| Outdoor and remaining items (see Fig. 17.8) | Codes 800 to 899 |

All job expenses are identified and coded as follows:

| Job expense (see Fig. 17.9) | Codes 900 to 999 |

The costs of all labor operations are identified and coded as follows:

| Labor (see Fig. 17.10) | Codes 000 to 099 |

A detailed discussion of cost codes follows in Chap. 17.

FORMS

Purchase Order Memo (POM) Form

This form is a four-part set which is used by either the coordinator or the storekeeper to initiate a purchase of materials. It serves as a draft copy which is used by the purchaser to make up the official purchase order. The POM form is illustrated in Fig. 16.2.

Purchase Order (PO) Form

This form is a seven-part set which is made up by the purchaser and approved by the project manager and the owner. The PO form is illustrated in Fig. 16.3.

Receiving Report (RR) Form

This form is a three-part set which is used to record the receipt of materials on the jobsite. The materials received are checked against the supplier's packing slip and the covering purchase order. The RR form is illustrated in Fig. 12.2.

Material Requisition (MR) Form

This form is a three-part set which is used by the field foreman to requisition materials from the stores. It is illustrated in Fig. 16.4.

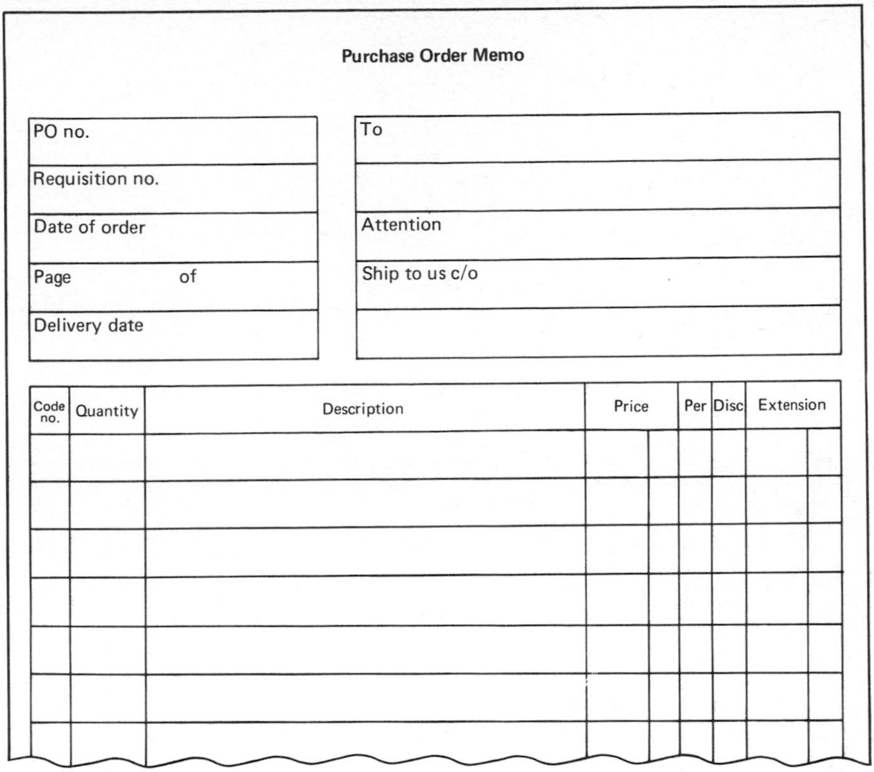

Purchase Order Memo

PO no.	To
Requisition no.	
Date of order	Attention
Page of	Ship to us c/o
Delivery date	

Code no.	Quantity	Description	Price	Per	Disc	Extension

FIGURE 16.2 Purchase order memo form.

Material or Tool Return/Transfer (MR/T) Report

This form is a four-part set which is used by the job stores to record the return or transfer of materials or tools to a designated location offsite. It is illustrated in Fig. 12.3.

Work Order (WO)

This form is a four-part set which is used to authorize the work outlined therein to proceed. It must be signed by a person who has the authority to do so and is illustrated in Fig. 15.7.

Voucher

This form is a four-part set and is a record of the material and time incorporated in the work as authorized by the covering work order. This form is illustrated in Fig. 15.8.

Purchase Order

PO no._____

To

Mark this number on all invoices, packing
slips, correspondence, etc.

Date of order	
Page of	Attention
Authorized signature	
	Ship to us c/o
Tender invoices in copies	
Sales tax included unless otherwise stated	Via f.o.b.
	Delivery required

Code no.	Quantity	Description	Price	Per	Disc.	Extension

CONFIRMATION

Confirmation only   This order subject to terms and conditions on reverse side

FIGURE 16.3   Purchase order form.

Daily Labor Report

This is a one-part form which is used to record the worker-hours expended daily by each worker, broken down into codes to designate the
operation and the area worked in. It is made up daily by the foreman
and is illustrated in Fig. 16.5.

Weekly Labor Distribution Form

This three-part set is the time sheet. It records the total hours spent on
the job by each worker during the week for payroll purposes. The form
is illustrated in Fig. 15.1.

Requisition for Material from Stores

Charge				Date		
Quantity		Description			Unit cost	Amount

| Requisition issued by (foreman's signature) | | | Filled by | | | |

FIGURE 16.4 Requisition for materials from stores form.

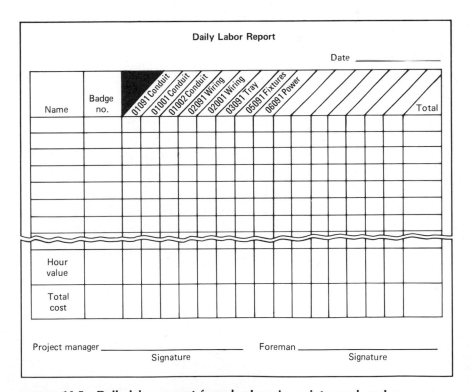

FIGURE 16.5 Daily labor report form broken down into work codes.

Field Force Report

This is a one-part form which is used by the field to record the number of workers on the site during every working day of the month. It is illustrated in Fig. 16.6.

Lost Tool Report

This form is a three-part set which is used by the field to report in detail the loss of tools. It is shown in Fig. 16.7.

Tool Record

This form is a two-part set, which is used for the monthly tool inventory. It is illustrated in Fig. 16.8.

Debit Memo

This form is a five-part set which is used to correct errors in a supplier's invoice. It is illustrated in Fig. 16.9.

Field Force

Month and year																Day															Remarks
	1	2	3	4	5	6	7	8	9	10	11	12	13	14	15	16	17	18	19	20	21	22	23	24	25	26	27	28	29	30	31	
Superintendent																																	
General foremen																																	
Foremen																																	
Electricians																																	
Apprentices																																	
Stores																																	
Others																																	
Total field force*																																	
Absentees†																																	
Worker Compensation Board†																																	

*Total includes all personnel carried on payroll.
†Counted in total field force, but listed separately for additional information.

FIGURE 16.6   Field force report form.

227

Lost Tool Report

Date _____

Tool no. _____ Serial no. _____

Manufacturer _____

Catalog no. _____ PO no. _____

Tool description _____

_____

Date tool last accounted for _____

Date tool last out to site on _____

Tool out to (worker's name) _____

Badge no. _____

Foreman's name _____

Remarks _____

_____

_____

_____

Stores coordinator _____
                              Signature

_____

For office use only

Cost of tool $ _____ Date of purchase _____

Delivery date _____ MR no. _____

Supplier _____

Does tool require replacement?   Yes _____ No_____

New tool replaced on PO no. _____

At a cost of $_____ ATIP

Project manager _____
                            Signature

FIGURE 16.7   Lost tool report.

Tool Record List

Date _____

Quantity	Description	Serial no.	Tool no.	Date received	Delivery slip no.	Checked by

FIGURE 16.8   Tool record list.

Debit Note	Debit to						
	Attention						

Date	PO no.			Debit no.			
Invoice no.		Amount		Your reference no.			
Quantity	Description			Price	Per	Disc.	Extension

DEBIT

Note: If credit to cover is not received within 10 days, this amount will be deducted from our next remittance.

FIGURE 16.9   Debit memo form.

230

17
Managing and Controlling with Computers

ROLE OF COMPUTERS

As you struggle to manage and control your jobs, you are, no doubt, deluged with an ever-increasing volume of paper work. This is the necessary evil that construction people dread and too often neglect.

Poor paper work, which is often a symptom and indication of poor communication between management and the jobs, is an important cause of low labor productivity and low job profitability. Any facility, therefore, that will reduce the volume of paper work, yet increase and speed up the access to vital information, is one that should command your interest. With the advent of minicomputers this facility came within the reach of many contractors. Now, with microcomputers, this facility is within the reach of every contractor.

Many construction people are frightened by computers or feel threatened by them. As a start, therefore, you must overcome the feeling that computers are some form of magic and realize that they are, in fact, sophisticated tools like many other sophisticated tools which you use and have grown quite comfortable with. A computer is a calculating machine with a TV screen. It can store and retrieve information and do complicated calculations almost instantly. The information thus obtained will be accurate and useful, provided the computer is properly programmed, and provided accurate and pertinent information is fed into it. That is the big problem. You are all aware of contractors who jumped into computer use with inadequate planning and preparation, half-baked programs, and untrained personnel and had disastrous results.

These contractors had unreasonable expectations that their computers would solve all their problems. They did not realize that computers are like new employees. Time and money must be spent on them to teach them the business and to teach them the ropes. How can you teach the computer your business, unless you have first undertaken the very difficult task of analyzing your existing operations and determined what are the problems and what exactly you are trying to accomplish?

The past history of computers in the construction industry has been far from satisfactory. After being seduced by the sales pitches and exciting promises of computer salespeople, many contractors had terrible letdowns when the resulting problems engulfed them. They had allowed computer programmers, who had neither gut knowledge of nor in-depth experience in construction, to sell them "canned" programs that very often did not satisfy their requirements. This was not only the fault of the computer industry. Some responsibility must rest with the contractors and their associations, who realized too late that they had to get more intimately involved with the experts if they wanted to produce satisfactory software programs that used the potential of the computer to really help them run their business.

Computers can and will play an important role in your business because of their ability to digest large quantities of information and to perform many detailed calculations accurately and quickly. Well-planned, straightforward management programs are required that assimilate and summarize the daily information from the head office and from the jobsites, and make it available easily and rapidly to key personnel. With fast, accurate, updated information—such as job status, job progress, job cost, productivity analysis, material status, tool status, and completion forecast—key personnel will be better able to perform their functions of management and control.

MANAGEMENT AND CONTROL FUNCTIONS

There are four major categories of management and control functions.

1. Accounting management
   a. Payroll
   b. Job costs
   c. Job labor cost
   d. Accounts receivable/payable
2. Financial management
   a. Cash flow
   b. Profit-loss analysis

c. Forecasting and budgetary planning
3. Project management
    a. Estimating
    b. Document control
    c. Production planning and scheduling
    d. Labor productivity analysis
    e. Change-order analysis
    f. Impact cost analysis
    g. Market sales and distribution analysis
4. Material management
    a. Purchasing and expediting
    b. Inventory control
    c. Tool and equipment cost analysis
    d. Tool and equipment control

The functions listed under the above four major management headings are performed to a lesser or greater degree by all contractors. These functions, most of which are done manually at present, require a tremendous amount of paper work and filing. Of the four management categories, only the first is now done by computers with any degree of sophistication. If you have a computer, it is probably performing your category 1 functions.

TEACHING THE COMPUTER

Everyday, as you go about your business, you are generating information which somehow must be written down, stored, retrieved, and communicated to the right parties if you are to properly perform your management functions. Procedures to achieve this should flow naturally and easily from the normal manner in which you perform your work. Filling out complicated reports, searching for information or files, and making sure that important information is communicated to the right parties at the right time are all tasks that do not flow naturally and easily in your everyday work patterns. They are, therefore, too often neglected or omitted.

Every management system is based on proper cost records and progress reports. Without them, you have management by crisis. Most contractors prepare some sort of schedule to manage their work, but less than half of them update it or keep it up to date in a continuous and meaningful manner. Most contractors maintain some form of cost records. Very few, however, are in sufficient detail to give them the current financial status of any given project or to forecast the final cost. Probably fewer than one-third of the contractors maintain proper cost records. Most contractors get some form of job progress reports, but very few relate these to quantities of work done. It is evident that there is a lot of room for improvement.

The advent of microcomputers may be a blessing for your industry. Because they are economical and within the means of most contractors, they should spark an in-depth review of your management procedures. In order to teach your business to your computer, you will have to analyze every operation and break each one down into step-by-step activities and decisions. You will have to understand these precisely and in detail, since you will never be able to teach the computer to make judgments or allowances based on a gut reaction to a situation. Your attitude towards using a computer to help manage your business will depend on the size of your company. Your reaction will probably depend on which of the following three groups you fit into.

1. Small shop employing up to 10 workers, managed and controlled by the owner. These one-person operators carry most of the management and control functions in their heads and can get along with a minimum of records.

2. Medium-size contractor employing up to 100 workers. These contractors must depend on staff to carry out the required management and control functions. Usually their budgets limit the size and quality of the staff and of systems necessary to produce records which are reliable and more than minimal.

3. Large contractors employing more than 100 workers. These contractors usually have sophisticated management systems and record keeping. However, they very often suffer from inadequate staff, unreliable records, and trial-and-error control techniques. They are probably already using computers, but mostly for payroll and accounting purposes.

The one-person operators are under no special pressure to get into computers, other than to jump on the bandwagon with a new tool that might increase their capacity to perform. The medium-size contractors are the ones who would probably benefit most from the use of microcomputers, provided that they are able to obtain the right software, that is, the right programs to suit their needs—not only accounting programs, but also project and material management programs. The larger contractors, as well, need to expand their existing computer programs to more fully service the needs of their project staffs, particularly in the areas of project management, productivity control, and the handling of information.

A COMPUTER ON EVERY JOBSITE

The one area where computers have not as yet penetrated to any great extent is on the jobsites themselves. Microcomputers are now in the same price range as many other sophisticated tools that can be found on most of the larger jobsites. The microcomputer can become an indispensable tool for project managers and their foremen. These job computers can be programmed so that the project supervisory personnel can call up important data at the touch of a key.

The following are the types of questions for which a job requires fast answers:

- What is it?
- Where does it go, and how?
- What can be done to reduce the cost?
- When will the materials arrive on site?
- How many worker-hours are required for the work?
- How many worker-hours were actually spent?

The systems for obtaining answers to these types of questions are easily programmed into a microcomputer and are as follows:

1. A summary of the specifications and the pertinent notes listed on drawings and in the estimate.
2. A cross-reference between operations and drawing numbers.
3. A description of applicable fastening methods.
4. A description of possible problems to look for.
5. Information regarding purchase orders, promised deliveries, back orders, and names of suppliers, including the people to contact and their phone numbers.
6. A labor-management program based on productivity analysis, which will provide the following data: worker-hours for a given area, worker-hours for a given operation, and worker-hours earned versus worker-hours spent. In this way, worker-hours can be targeted, crew sizes established, and productivity checked.
7. A bar chart which compares estimated and actual performance. A picture is worth a thousand words, and high-quality graphics are

available as a software option. Graphics are an efficient way of conveying a large amount of information, particularly detailed numerical data, quickly and simply to your job staff. Bar charts and other graphs can better help them manage the labor and thus improve productivity.

The computer won't think for project managers, or make difficult decisions for them, or tell them what must be done. It can, however, supply relevant data to help them perform these functions better. Construction is undisciplined, and every job is a make-up team. Project managers need all the support that they can get.

It is possible almost to eliminate the waste of worker-hours spent on phone calls and searching through files, and to eliminate the many mistakes which result from the inability to obtain required information on time. How? By putting a microcomputer on the site and programming it with information that is readily available from the estimate, the purchase orders, the specifications, and historical data garnered from similar jobs.

The first step is to train the project staffs to be familiar and comfortable with their microcomputers and to accept them as necessary and indispensable tools. The economics are right; the software is straightforward and is available to implement this first step now. The impact on jobs will be felt quickly and will result in the rapid improvement of project management and control. After microcomputers have become necessary tools on jobsites, in the same manner as telephones and copiers, you may be ready to further expand this management facility.

In the next phase—which no doubt is still a few years in the future —computers or terminals on jobsites can be tied in to a central computer at the head office. In this phase, the everyday information generated by the key personnel at the head office and the jobsites will be fed into the computer terminals where the work is being done, rather than into memos or reports, and will be filed in the central computer, rather than in filing cabinets. Key people in the firm will have access to the pertinent data, by calling up on their terminals the information stored in the computer. These key people will also feed into their terminals (in a prescribed manner) information which they generate and are responsible for, so that this information will become part of the data bank of the central computer. The central computer can be programmed to digest, sort out, compare, forecast, and store this information for instant retrieval by the key personnel who require it in order to properly manage and control your business.

Because information can be fed in continuously, as it happens, the required information, available on the terminal screens, will be up-to-date and pertinent. A printer can be built in with the central computer at the head office, so that any information can be printed out if required. This may sound farfetched, but the technology is available to make it work—at an affordable price, if one considers the large sums of money that contractors spend on performing these functions manually, or the losses which result when these functions are not performed.

The problem, as stated before, is to develop programs that are straightforward and that flow naturally and simply from the day-to-day activities of the people who generate the information. These people, in turn, must get back information that is pertinent and will help them manage and control their activities. This is easier said than done. Like the cashless society, which is beginning to develop but is still far from a reality, a paperless, instantly communicating, computer-based construction industry is still in the future. Nonetheless, many of the management functions previously listed can now be performed by using a relatively affordable microcomputer, with programs which have been sufficiently developed and refined to be within the scope of most contractors.

Before embarking on a computer program, contractors must first analyze and perfect their existing, mostly manual, systems. Many of these have developed spontaneously, and are overlapping and unclear, or they may be completely outdated. Only if you know what you want and are able to derive it from your existing systems in a clear, efficient manner will you be able to effectively change over to a computer.

INPUT—OPERATION REPORTS

The foundation of an effective reporting system for managing and controlling your jobs rests on the following data systems:

1. The estimate, and how you use it to plan and control your job budgeting and forecasts, cash flow projections, profit projections, etc.
2. The time sheet system, and how you use it for payroll, labor cost control, productivity analysis, budgeting and forecasts, etc.
3. The purchase order system, and how you use it for material cost control, budgeting and forecasts, cash flow analysis, etc.
4. The job progress report system, and how you use it to obtain an accurate picture of what has been accomplished on the job, how much was consumed, and how long it took

Since the estimate is the vehicle that very often carries you to the job awarded to you, it must also be the vehicle for planning the job, targeting the various operations, and establishing their respective productivity requirements. The estimate must be your bible, and your field people must refer to it religiously. If your operating personnel really understand that, in a firm price contract, what you have in your estimate is what you have for doing the job, then they will use it to plan and manage the job so that they will meet their targets.

The time sheet is probably one of the most accurate, most carefully scrutinized documents in your business. By coding it so that it shows not only how many hours each person worked, but how many hours were worked at given operations, it becomes a valuable costing and control tool. Particularly in certain operations, if the number of hours worked can be related to the quantity of material installed, then you have a field check of productivity.

The purchase order is similar to the time sheet: properly coded, it is a valuable tool for costing and cash-flow projections. Together with a quantity control of designated critical materials, such as conduit and wire, it will give you a valuable insight into what is happening on the job in time to take corrective actions, if necessary.

The progress report system must be tightened up to reflect accurately what is happening on the job. Reports tend to be optimistic at the beginning, and by the time the real facts become indisputably clear, it may be too late to turn things around. So the reporting system must be based not only on judgments of work completed, but also on actual quantities installed. It is probably not necessary to do this on every operation, just on a few critical ones that will give you a handle in controlling the job. Critical operations are those which control the overall timing of the job. Only a small percentage of the operations on a project are critical. Control them, and you will control your job.

As far as data processing is concerned, anything you can do, the computer can do faster and better—anything, that is, except think, plan, and conceptualize. You have to supply these functions and inputs. You must realize that you are the central feature of the computer system, and that a computer is just a very sophisticated tool whose output will only be effective and useful if your input is well thought out, disciplined, and programmed to suit your needs. As with every sophisticated tool, your people must be properly trained to use it.

OUTPUT—STATUS REPORTS

By utilizing the four reporting systems previously described and by coding the cost of materials, job expenses, and the cost of labor, you can obtain the following vital information:

1. Budget status
   a. Original budget
   b. Approved change orders

    c. Current budget

    d. Pending change orders

    e. Forecast final budget

2. Payment versus cost status

    a. Payment to date

    b. Total cost to date

    c. Forecast payment to complete

    d. Forecast cost to complete

3. Schedule status

    a. Planned start and finish and percent complete

    b. Current start and finish and percent complete

    c. Percentage of work done

    d. Percentage of total budget spent

    e. Performance ratio

4. Labor status

    a. Original labor budget

    b. Revised budget (including changes)

    c. Earned this period

    d. Actual cost this period

    e. Earned to date

    f. Actual cost to date

    g. Forecast to complete

    h. Variance from revised budget

5. Labor status per cost code

    a. Original budget

    b. Revised budget

    c. Scheduled start and finish

    d. Percent complete

    e. Worker-hours earned

    f. Worker-hours actually spent

    g. Percentage of budget spent

    h. Performance to date, percent

    i. Forecast performance, percent

6. Material status

    a. Original material budget

    b. Revised budget

    c. Purchase orders committed

    d. Invoiced to date

    e. Forecast cost to complete

    f. Variance

  7. Material status per cost code

    a. Original budget

    b. Revised budget

    c. Purchase orders committed

    d. Invoiced to date

    e. Forecast to complete

    f. Variance

  8. Quantity status per designated material code

    a. Estimated total quantity

    b. Revised total quantity

    c. Total this period

    d. Total to date

    e. Forecast to complete

    f. Variance

  9. Productivity status per designated labor operation

    a. Estimated unit worker-hours

    b. Actual installed unit worker-hours

    c. Productivity performance, percent

 10. Planning and resource control status

    a. Personnel: demand versus availability (based on 5-day week, single shift)

    b. Scheduled delivery of material

    c. Promised delivery of material

CODING SYSTEM

A coding system is used to break down the total materials, job expense, and labor of your contract into smaller, identifiable groupings in order to be able to manage and control them to meet your targets. Too many pieces will make the system unwieldy and unmanageable. You will not be able to see the forest for the trees. Too few pieces will not give you sufficient control. The manner in which you break down the materials and labor should reflect the way in which you do your purchasing and your installation in the field. The simplest system allowing you to accomplish this, while providing an acceptable level of control, is the best.

The following is a simple coding system for materials, job expense, and labor. The basic identification is a three-digit code number. The first digit is the code control, which identifies the general category. The next two digits are the item control, which identifies a particular item in the general category.

There are eight general categories for materials, one general category for job expense, and one general category for labor. Each general category can be broken down into 100 items. In the codes that follow, suggested minimum numbers of items are utilized. You can expand the proposed codes by breaking down the general categories into many more items, up to their full capacity each of 100 items, giving a total capacity for this coding system of 1000 items.

The ten codes shown in Figs. 17.1 through 17.10 are suggested as a guide. You can, of course, change the listings and increase the quantity of items to suit your particular needs.

Material Codes

The code numbers for the general categories of materials are shown in Table 17.1.

A general code number, for example, code 100, covers the total quantity and cost of all the materials in that general code. This general code number can be subdivided into 99 itemized codes, if desired, that is, 101, 102, 103, etc. Each of the itemized code numbers isolates and

**TABLE 17.1   CODE NUMBERS FOR
GENERAL CATEGORIES OF MATERIALS**

CODE NUMBER	MATERIAL CATEGORY	ITEM NUMBERS
100	Conduit	100–199
200	Wire and cable	200–299
300	Raceways and nonmetallic pipe	300–399
400	Accessories and finishing materials	400–499
500	Fixtures and lamps	500–599
600	Power and distribution equipment	600–699
700	Systems equipment	700–799
800	Outdoor and remaining materials	800–899

identifies a particular material or group of materials which form part of the general category, and allows you to obtain a separate printout or a separate costing for this code number. Thus item 110 pertains to rigid conduit, ½ to 1 in, as shown in the coding list in Fig. 17.1. The subdivisions have been kept to a bare minimum for functional control purposes. You can, of course, subdivide them further, to their full capacity, to suit your own requirements.

Job Expense Codes

The general code number for direct job expense is 900. The suggested subdivision of this general category covers only the minimum number of items usually encountered on a job. You can break this general category into many more items (up to 99) to suit your requirements. The suggested subdivision shown in the coding list (Fig. 17.9) is as follows.

1. Shopwork materials
2. Warehouse and transfer materials
3. Subcontract work
4. Work-order materials
5. Rentals of tools, hoisting equipment, etc.
6. Job office expenses
7. Drawing and printing expense
8. Independent testing
9. Bonding and financing charges
10. Room and board, traveling expenses
11. Permit and inspection fees

Labor Codes

The general code number for job payroll labor is 000. This general category is subdivided into nine control headings which cover the main types of work performed on a typical contract. Each of these main headings can be subdivided into ten items to isolate the specific operations that you may want to keep track of. The breakdowns should support your cost control and productivity analysis program.

The suggested breakdown of the general labor code is shown in Fig. 17.10. In the figure, finishing work (items 040 to 049) covers the installation of light switches, receptacles, plates, connection of motors, testing, and any other accessory or finishing operations. Outdoor work and remaining equipment installation (items 080 to 089) pertain to the

CODE NO. 100: CONDUIT	MATERIAL CODE NUMBER	
	CATEGORY CONTROL	ITEM CONTROL
Total material under this code number	1	00
Rigid conduit, ½ to 1 in	1	10
Rigid conduit, 1¼ in and up	1	11
Rigid conduit accessories	1	12
EMT conduit, ½ to 1 in	1	20
EMT conduit, 1¼ in and up	1	21
EMT conduit accessories	1	22
Flexible conduit	1	30
Flexible conduit accessories	1	31
Plastic-coated rigid conduit	1	40
Plastic-coated rigid conduit accessories	1	41
	1	50
	1	60
	1	70
	1	80
	1	90

FIGURE 17.1   Code no. 100: conduit.

CODE NO. 200: WIRE AND CABLE	MATERIAL CODE NUMBER	
	CATEGORY CONTROL	ITEM CONTROL
Total material under this code number	2	00
RW90, TWH up to no. 10	2	10
RW90, TWH no. 8 and up	2	20
BX, aluminum-sheathed, nonmetallic-sheathed	2	30
Teck cable (armored cable)	2	40
High-voltage cable	2	50
Systems cable	2	60
Special cables	2	70
	2	80
	2	90

FIGURE 17.2   Code no. 200: wire and cable.

CODE NO. 300: RACEWAYS AND NONMETALLIC PIPING	MATERIAL CODE NUMBER	
	CATEGORY CONTROL	ITEM CONTROL
Total material under this code number	3	00
Underfloor duct and fittings	3	10
Tray and accessories	3	20
Lay-in duct, wireways, etc.	3	30
Plastic conduit	3	40
Fiber/PVC/fiberglass ducts	3	50
	3	60
	3	70
	3	80
	3	90

FIGURE 17.3   Code no. 300: raceways and nonmetallic piping.

installation of any equipment or activities not covered in the other labor codes. Fabrication shop, work orders, etc. (items 090 to 099), cover labor costs that fall under the heading of job expense, such as

1. Shop labor to prefabricate assemblies required by the job
2. Labor expended on work orders
3. Labor for temporary wiring
4. Labor to demolish existing work
5. Labor lost due to impact factors beyond your control

The general labor code (000) allows you nine itemized codes to record vital general information that you may need to manage and control your contract. The following is a suggested list of uses for these nine item codes:

000	Total cost of labor, per the payroll
001	Cost of direct labor (that is, the portion of the job payroll paid to the workers in the field)
002	Cost of indirect labor (that is, the portion of the job payroll paid for supervision and administration)
003	Total worker-hours

CODE NO. 400: ACCESSORY AND FINISHING MATERIALS	MATERIAL CODE NUMBER	
	CATEGORY CONTROL	ITEM CONTROL
Total material under this code number	4	00
Outlet boxes and covers	4	10
Soc and junction boxes	4	15
Conduit accessories	4	20
Wire accessories	4	25
Teck connectors	4	30
Lugs, service connectors, terminal blocks	4	35
Flexible and seal-tight conduit accessories	4	40
Hub-type fittings	4	45
Explosion-proof fittings	4	46
Grounding accessories	4	47
Cadwelds, splicing materials	4	49
Switches, receptacles, plates	4	50
Fuses	4	60
Channel, angle iron, steel rods	4	70
Fasteners, anchors, inserts	4	75
Brackets, supports	4	80
Hardware, nuts, bolts, screws	4	82
Lumber, plywood, cement	4	84
Gases, oxygen, acetylene, propane	4	86
Pulling compound, cutting oil	4	88
Tape, solder, paints	4	90
Expendable tools, bits, taps, fishwire	4	95
Fireproofing	4	99

FIGURE 17.4   Code no. 400: accessory and finishing materials.

004   Direct labor worker-hours

005   Labor multiplier (total labor cost divided by number of direct worker-hours)

006   Premium time

007   Escalation

008   Cost of commissioning and testing

009   Cost of clean-up, corrections of deficiencies, and guarantee

CODE NO. 500: FIXTURES AND LAMPS	MATERIAL CODE NUMBER	
	CATEGORY CONTROL	ITEM CONTROL
Total material under this code number	5	00
Fluorescent fixtures	5	10
Incandescent fixtures	5	20
Mercury/sodium/quartz fixtures	5	30
Special fixtures	5	40
Fluorescent lamps	5	50
Incandescent lamps	5	60
Mercury/sodium/quartz lamps	5	70
Ballasts	5	80
Fixture accessories	5	90

FIGURE 17.5   Code no. 500: fixtures and lamps.

CODE NO. 600: POWER AND DISTRIBUTION EQUIPMENT	MATERIAL CODE NUMBER	
	CATEGORY CONTROL	ITEM CONTROL
Total material under this code number	6	00
High-voltage switchgear	6	10
Low-voltage switchboards	6	20
Power transformers	6	30
Bus duct and accessories	6	40
Panel boards and loose breakers	6	50
Dry-core transformers	6	60
Motor control centers	6	70
Starters, contactors, relays	6	75
Disconnect and transfer switches	6	80
Splitter, meter, control boxes	6	85
	6	90

FIGURE 17.6   Code no. 600: power and distribution equipment.

CODE NO. 700: SYSTEMS EQUIPMENT	MATERIAL CODE NUMBER	
	CATEGORY CONTROL	ITEM CONTROL
Total material under this code number	7	00
Fire alarm system	7	10
Public address system	7	15
Surveillance and security tour	7	20
Intercom system	7	25
Clock and program system	7	30
Radio system	7	35
Call system	7	40
Hospital Signaling systems	7	45
Dimming system	7	50
Emergency generator and battery system	7	55
Uninterruptible power system (UPS)	7	60
Lighting and energy control system	7	65
Electric heating	7	70
Power factor correction	7	75
Client metering	7	80
Closed-circuit TV	7	85
	7	90
	7	95

FIGURE 17.7 **Code no. 700: systems equipment.**

Location Code

The three-digit coding system may be sufficient to satisfy the requirements of many contractors. The potential of 1000 code numbers allows them to break down their contracts for adequate costing purposes. These codes are also easily applied during purchasing to segregate and identify the various types of materials. The three-digit codes can be further expanded by subdividing them into a larger quantity of subcodes to segregate the items according to their location on the jobsite.

By adding an additional two digits to the material or labor code number, you can identify the location where the material is going or where the work will be done. Take a 15-story office building with two basements as an example. The location code is shown in Table 17.2.

When the location code is 00, it refers to the total project. If you

CODE NO. 800: OUTDOOR AND REMAINING ITEMS	MATERIAL CODE NUMBER	
	CATEGORY CONTROL	ITEM CONTROL
Total material under this code number	8	00
Outdoor electric tower	8	05
Outdoor duct banks	8	10
Pole line	8	15
Light standards	8	20
Motors	8	25
Appliances, fans, etc.	8	30
	8	40
	8	50
	8	60
	8	70
	8	80
Temporary wiring	8	90

FIGURE 17.8 Code no. 800: outdoor and remaining items.

CODE NO. 900: JOB EXPENSE ITEMS	JOB EXPENSE CODE NUMBER	
	CATEGORY CONTROL	ITEM CONTROL
Total job expense under this code number	9	00
Shopwork materials	9	10
Warehouse and transfer materials	9	20
Subcontract work	9	30
Work orders	9	40
Rentals (tools, hoisting, cranes)	9	50
Job office (equipment, stationery, telephone)	9	60
Drawings, printing	9	70
Independent testing	9	80
Bonding	9	85
Room and board, travel	9	90
Permits and fees	9	95
	9	96
	9	97
	9	98
	9	99

FIGURE 17.9 Code no. 900: job expense items.

CODE NO. 000: LABOR	LABOR CODE NUMBER	
	CATEGORY CONTROL	ITEM CONTROL
Total labor under this code number	0	00
Conduit work	0	10
Wire and cable work	0	20
Raceways and nonmetallic pipe	0	30
Finishing work	0	40
Fixtures and lamps	0	50
Power and distribution equipment	0	60
Systems installation	0	70
Remaining equipment or operations	0	80
Fabrication shop, work orders, temporary wiring, etc.	0	90

FIGURE 17.10   Code no. 000: labor.

wish to code the supply of 2-in EMT conduit for the first basement, the code number is 12191. If you wish to code the labor to install the conduit on the twelfth floor, the code number is 01012. Thus, with the addition of the two digits to identify the location, you can code the materials during purchasing so that they will be identified as to their point of installation on the jobsite. This will facilitate the material handling and make this activity more efficient. The expanded material location codes and labor location codes are used to further break down the material codes and labor codes for more detailed identification and costing purposes. This will allow you to fine-tune your control procedures.

A labor code, when expanded by the location codes, gives you the labor subcodes. For want of a better name, you can call these the work codes. The work codes identify where the work is done. When utilized in your costing system, each work code identifies the portion of the cost of the labor code which is expended in that given location. The labor codes have a capacity of 100 items, as previously shown. The location codes have a capacity of 100 items. Therefore, you have a potential of 10,000 work code items. For example, work code 01012 identifies the installation of conduit on the twelfth floor. When this work code is used in your costing system, it identifies the cost of the labor to install the conduit on the twelfth floor. Work code 01000 identifies the labor to install the conduit in the total project. When used in your costing system, this work code identifies the cost of the labor to install all the

TABLE 17.2 LOCATION CODES FOR A FIFTEEN-STORY BUILDING

FLOOR NUMBER	LOCATION CODE
Basement 2	92
Basement 1	91
Ground	01
2	02
3	03
4	04
5	05
6	06
7	07
8	08
9	09
10	10
11	11
12	12
13	13
14	14
15	15

conduit throughout the entire project. In your costing system, the work codes are your labor cost codes.

Since it is unlikely that you will use all the location numbers to identify the different areas that you work in, some of the unused location numbers can be utilized to identify specific operations that you may wish to keep track of. If, for example, you are confronted with a start-and-stop pattern of work in the installation of trays because of excessive changes and delays in the structural work, you may wish to cost this operation to support a future claim for lost time. You can set up a work code for this purpose by using the labor code for trays, 030, and adding on one of the unused location numbers. This becomes your work code for the actual labor to install the trays under the unfavorable conditions. By comparing this actual labor with either your labor for doing similar work under normal conditions or your estimate, you can establish the amount of lost time.

A material code when expanded by the location codes gives you the material subcodes. The material subcodes identify the portions of the material code that are installed at each given location. When utilized in your costing system, the material subcodes identify the cost of the items of the material code that pertain to the given location. The material codes have a capacity of 800 items. The location codes have a capacity of 100 items. You, therefore, have a potential total of 80,000 material subcode items. In your purchasing system, these material subcodes are utilized for identification and checking purposes. When they are utilized in your costing system, they become material cost codes.

Since it is unlikely that you will use all the location numbers to subdivide the material codes into material subcodes, some of the unused location numbers can be used to identify specific material cost breakdowns that you may wish to keep track of. For example, you may wish to identify the cost of materials on a large extra on which you are working, the price of which has not been settled.

The purpose, again, is to keep the coding system as simple as possible, yet to use it in a flexible and effective manner. However you use it, you must be consistent, so that your personnel will become familiar with the codes and their use will become habitual to them. Individual work codes should be sufficiently distinguishable from other project activities so that the coding will allow for this work to be readily measured. The same applies to materials.

By applying the location codes to the job expense codes, you can also expand these into job expense subcodes. You may wish to do this in order to segregate the job expenses and allot or relate them to various areas or operations. The breakdown of the job expense, the labor, and the material into subcodes may be particularly useful as a backup for your progress billing, as will be described in the following chapter.

18
Progress Billing and Cash Flow

ENTITLEMENT TO PAYMENT FOR PERFORMANCE OF WORK

As you perform your work on a contract, you are entitled to be paid on a regular basis in order to meet your financial commitments to your staff, workers, suppliers, and other creditors. You request payment by means of progress billings on a regular basis. In most cases these requests are submitted on or about the twentieth of every month.

The record in your industry, regarding payments to contractors in a reasonable period of time, is not that great. The money passes through many levels and many hands, and there are many factors that tend to slow down the process. Some of these are legitimate, some are illegitimate, and some are due to sloppiness or lack of proper communication.

Your progress billing has to be approved, usually by the architect or the engineer. Most contracts require some form of breakdown of your overall price, to which you are expected to apply percentages based on the work completed. Needless to say, this very often leads to differences of opinion. The result may be that your progress billing is cut, sometimes substantially, below the amount that you have requested. In your billing you try to, at least, cover your cost. The architects and engineers are not necessarily concerned with your cost. They are only prepared to certify what they think is the value of the actual work put in place. The amount that they will approve will be influenced by how factually you can explain what you have performed and by the manner in which you have presented your progress billing.

To arrive at a responsible judgment of the percentage completed of an operation can be very tricky. Many engineers, and construction people as well, tend to focus on the actual installation activity itself as representing the cost for billing purposes. As was pointed out in Chap. 2, the installation activity consumes only a part of the total cost to do the work. Another part of the total cost is due to worker-hours expended on planning, coordination, and material handling. Thus, even before any of the actual installation work commences on any operation or in any area, you may already have expended considerable worker-hours on planning and coordinating the work and on material storage and handling. The costs for these activities are included in your price, and you are entitled to bill for them when they occur, rather than after the installation activity is completed. This applies, as well, to your job expense, general overhead, and profit, which you are entitled to bill out in relation to the duration of the project and the percentage of the contract completed.

It is desirable and expedient to work out a procedure that will expedite approval of your monthly billing and thus speed up your cash flow. Developing such a procedure is imperative in these times of inflation and high interest rates. You have to understand the time value of money. When your progress payments are reduced or delayed, you must pay substantial amounts for financing costs to cover the shortfall. It is in your vital interest to develop a billing procedure that will speed up approval.

BREAKDOWN OF THE CONTRACT FOR BILLING

In order to expedite the billing procedure and subsequent approval time, arrange to break down your contract into a sufficient number of material codes and labor codes to present a practical listing for rapid checking purposes. Utilize the form shown in Fig. 18.1 for this purpose. In column 1, list the applicable code numbers. In column 2, write in a brief description of the respective code numbers. In column 3, list the amounts for the respective codes. The total of all the amounts in column 3 adds up to your contract amount. In column 4, list the various location codes in which the work will be done. The amount of each code number is subdivided into a subtotal for each location. This gives you a listing of the material cost codes for each material code, a listing of the expense cost codes for the job expense code, and a listing of the work codes for each labor code. This is your master sheet, which will only change if the code amounts are revised because of change orders added to the contract. In some contracts the change orders are listed separately and billed separately.

(1) Code no.	(2) Description	(3) Amount of code	(4) Breakdown of code amount per location (cost codes)								
			91	01	02	03	04	05			
110	Rigid conduit, all sizes	10,000	4,000	1,000	1,000	1,000	1,000	2,000			
120	EMT conduit, all sizes	15,000	3,000	2,500	2,500	2,500	2,500	2,000			
220	Wire, all sizes	22,000	6,000	3,000	3,000	3,000	3,000	4,000			
230	BX	12,000		3,000	3,000	3,000	3,000				
240	Teck cable and HV cable	28,000	18,000					10,000			
400	Accessories	60,000	12,000	10,000	10,000	10,000	10,000	8,000			
500	Fixtures and lamps	115,000	10,000	20,000	20,000	20,000	20,000	5,000			
610	Switchgear substations and power transformers	310,000	210,000					110,000			
640	Bus duct, panels, dry transformers	140,000	60,000	10,000	10,000	10,000	10,000	40,000			
710	Fire alarm and PA systems	40,000	6,000	6,000	6,000	6,000	6,000	10,000			
755	Emergency generator	136,000						136,000			
760	Uninterruptible power supply	224,000						224,000			
800	Balance materials	100,000	25,000	15,000	15,000	15,000	15,000	15,000			
900	Job expense and administration	102,000	22,000	17,000	17,000	17,000	17,000	12,000			
Labor											
010	Conduit work	110,000	28,000	18,000	18,000	18,000	18,000	10,000			
020	Wire and cable work	75,000	16,000	13,000	13,000	13,000	13,000	7,000			
040	Finishing work	30,000	5,000	5,000	5,000	5,000	5,000	5,000			
050	Fixture and lamp installation	75,000	10,000	15,000	15,000	15,000	15,000	5,000			
060	Power and distribution equipment installation	60,000	10,000	8,000	8,000	8,000	8,000	18,000			
070	Systems installation	46,000	6,000	5,000	5,000	5,000	5,000	20,000			
080	Remaining equipment installation	10,000	1,000	2,000	2,000	2,000	2,000	1,000			
090	Fabrication shop	40,000	4,000	8,000	8,000	8,000	8,000	4,000			
094	Temporary wiring work	25,000	6,000	4,000	4,000	4,000	4,000	3,000			
	Total contract	1,785,000									

FIGURE 18.1 Breakdown of the contract into cost codes for progress billing.

PERCENTAGE COMPLETED FOR BILLING

The progress of the work for billing purposes is tabulated on the percentage completed form, illustrated in Fig. 18.2. The code numbers are repeated in column 1 of this form. Around the twentieth of every month, the project manager lists the percentages completed to date of every active material cost code, expense cost code, and work code and reviews these with the architect or engineer in order to obtain approval. The percentage completed form with the approved percentages listed therein is sent to the office.

At the office the percentages are applied to the respective cost codes, and the amount expended to date for each active code is listed in column 3. The amount of each active code which appeared in column 3 of the previous billing is posted in column 4. The difference between column 3 and column 4 represents the amount expended this month for every active code, and is posted in column 5. The percentage that the amount to date represents of its total code amount is listed in column 6. This gives you an indication of the completion status of each code and of the total contract.

PROGRESS BILLING SYSTEM

To illustrate this system of progress billing, refer to the data listed in Figs. 18.1 and 18.2, which refer to the breakdown of the electrical contract for a computer data center. The building consists of a basement, four typical floors, and a mechanical penthouse. The location codes are as follows:

Basement	91
Ground floor	01
Second floor	02
Third floor	03
Fourth floor	04
Mechanical penthouse	05

The number of codes into which the contract is broken down was discussed and agreed upon with the approval authorities. It was agreed to limit the codes to the minimum number that would still allow adequate checking for billing purposes. In this case there are 13 material codes, 1 expense code, and 9 labor codes. Each code amount is subdivided by location into cost subcodes. For example, code 110, which covers rigid conduit of all sizes, amounts to $10,000 and is subdivided

(1) Code no.	(2) Percentage completed per cost code						(3) Amount to date, dollars	(4) Amount previous, dollars	(5) Amount this month, dollars	(6) % complete to date
	91	01	02	03	04	05				
110	100	100	100	50	50	50	8,000	6,400	1,600	80
120	100	100	100	50	50		10,500	8,000	2,500	70
220	100	100	100	50			14,500	10,500	4,000	65
230		100	100				6,000	3,000	3,000	50
240	50					100	19,000	10,000	9,000	70
400	75	50	50	50	50	50	33,000	22,000	11,000	55
500	90	25	25	25	25	50	31,500	10,000	21,500	27
610	50					25	132,500	48,500	84,000	43
640	25	10	10	10	10	25	29,000	15,000	14,000	20
710	20	20	20	20	20	20	8,000	4,000	4,000	20
755						40	54,400		54,400	40
760										
800	50	20	20	20	20	30	29,000	20,000	9,000	29
900	25	25	25	25	25	25	25,500	22,000	3,500	25
Labor										
010	75	50	50	30	30	30	52,800	46,000	6,800	49
020	50	30	30	10	10	10	19,100	11,600	7,500	25
040	50	10	10	10	10	50	13,500	12,000	1,500	45
050	10	10	10	10	10	10	7,500	4,000	3,500	10
060	25	10	10	10	10	20	7,300	4,300	3,000	12
070	10	10	10	10	10	25	7,600	5,600	2,000	16
080	20	20	20	20	20	20	2,000	1,000	1,000	20
090	50	100	50			25	15,000	10,000	5,000	37
099	100	100	70	70	70	70	20,200	15,200	5,000	80
							545,900	289,100	256,800	30

FIGURE 18.2 Percentages completed for progress billing.

as shown in Table 18.1. This procedure is followed for all codes (see Fig. 18.1).

TABLE 18.1   SUBDIVISION OF MATERIAL CODE 110 INTO SUBCODES

MATERIAL SUBCODE		COST BREAKDOWN, DOLLARS
MATERIAL CODE	LOCATION CODE	
110	91	4,000
110	01	1,000
110	02	1,000
110	03	1,000
110	04	1,000
110	05	2,000
110	00	10,000

The approved percentages completed for each active subcode were listed by the project manager, and this enabled the office to work out the amount of each code completed to date. These amounts are listed in column 3 of the percentage completed form (Fig. 18.2). The total of column 3 adds up to $545,900. The corresponding amounts from the previous approved billing are listed in column 4 and add up to $289,100. The difference between column 3 and column 4 represents the amount completed during the billing month for each active category, and is listed in column 5. This column adds up to $256,800. Thus the monthly billing for the project in Fig. 18.2 works out as shown in Table 18.2.

There are enough factors acting to delay or reduce your progress billings without your adding to the problem. For this reason you should press for a procedure that will speed up the process. The advantage of a procedure such as the one just described is that it is based on a

TABLE 18.2   PROJECT MONTHLY BILLING

ITEM	AMOUNT, DOLLARS
Total billing to date	545,900
Billing last month	289,100
Billing this month	256,800
Less 10% holdback	25,680
Amount due this month	231,120

systematic manner of arriving at the billing amounts. It focuses on small enough segments of the work to allow for reasonable judgment and agreement on the percentage completed. It can also be easily programmed into a computer-based billing program.

The electrical contractor for the project itemized in Fig. 18.1 bid the job very competitively and was very concerned about the possible impact of financing costs on the profitability of this contract. The following analysis will illustrate what this contractor was up against. The prevailing conditions throughout the contract were as follows:

Contract price	$1,785,000
Contract duration	18 months
Holdback	10 percent
Inflation	10 percent
Prevailing interest rates	18 percent

The 10 percent holdback was deducted from every monthly payment, and the total holdback was paid to the contractor 12 months after substantial completion of the work. The contractor had to pay interest on the amount of the holdback at an annual rate of 18 percent, calculated on a monthly basis. The financing expense for the delayed holdback plus the devaluation due to inflation cost the contractor $35,-000. This is about 20 percent of the holdback amount. The $35,000 was over and above the regular financing expenses of the project which resulted from the actual cash-flow situation, and was in fact an additional cost suffered by the contractor because of the size of the holdback and the length of time it took to receive this money. This example illustrates what was meant by the foregoing statement that you have to be very conscious of the time value of money.

Expediting the cash flow in order to reduce the cost of financing requires a policy and an effort on your part. For example, mixed-up accounts delay payments. You, therefore, have to discipline your accounting department to regularly reconcile the contract amounts with the accounting department of the other party to your contract. You have to pursue progress payments diligently. If you don't press, you don't get. Otherwise, these payments and holdback releases will often be stalled for a variety of reasons or excuses. It is not uncommon for holdbacks and final payments to be delayed for six months or a year after the work has been completed. In such a case, the dollar that you earned and to which you are entitled may be worth only eighty cents by the time you finally get it.

ONEROUS PAYMENT CLAUSES

Contractor beware. You have to be very careful not to sign a contract with onerous payment clauses which will be utilized to delay the payments for work done. You have to read the fine print of the contracts, or else have your lawyer read it for you. Don't accept conditional clauses that will delay payment to you because of the actions of others or because of factors beyond your control. Don't accept any clauses that hold you liable for consequential damages. Such clauses can hold you responsible for damages for intangibles, such as lost profits, lost business, lost savings, or any other similar type of loss.

Slow payment or difficulty in collecting can sometimes be an early warning signal that the party with whom you signed the contract is in financial trouble. The best protection that you have to ensure payment for your work is your *lien rights*. You should be aware of these rights and how to use them.

LIENS AND PRIVILEGES

When you have put material and work into a building or, as it is legally called, an immovable, the law gives you certain preferred rights as a creditor. These are commonly known as lien rights or privileges.

The protection is not automatic. Some awareness and action are required on your part. First, you have to ascertain, even to the extent of checking at the registry office, who is the owner of the project, and that there is nothing registered that will interfere with your lien rights. Then you have to notify the owner, in writing, that you have a contract for a given portion of the work at a given price.

On some contracts, you may be asked to sign a waiver or renunciation of your lien rights. Obviously, the best advice is not to sign such a clause. However, in some cases it may be mandatory if you are to get the contract. On some projects, the lenders or mortgage companies insist on obtaining waivers from the contractor and subcontractors before they will release the monies for the project. In most such cases, the waiver gives the lender first rights to recovery, but you can still register your privilege for recovery from what is left. Before signing a waiver clause, you would do well to check it out with your legal advisor.

A privilege or lien must be registered a prescribed number of days before the building is substantially completed. In practice this is not as simple a step as it may seem, since contractors have been conditioned to take many knocks before they accept the inevitable. Registering a lien is a last resort, and the onus is on you to make sure that your action is valid and that you registered the lien on time. If you have registered a valid lien and are confronted with a bankruptcy, then you become a secured creditor. Otherwise, you become an ordinary creditor, and your chances of recovery are very slim.

A bankruptcy results in a meeting of the creditors, at which a trustee is appointed or confirmed, along with inspectors, who are usually the largest creditors. The trustee, with the help of the inspectors, administers the estate. Trustees are usually chartered public accountants who are licensed and bonded to perform this function. You have the right to challenge the work of the trustee and inspectors if you feel that their actions are not adequate to protect your interests, or if you feel they are not performing their functions properly.

Early in the proceedings, you will be required to file your proof of claim. The bankrupt may make a proposal, which can only be accepted if the majority of the creditors and those holding 75 percent of the value of the claims vote to accept it. Alternatively, the trustee will dispose of the assets of the estate in such a manner as will effect the largest possible return. The secured creditors will be looked after first, and if there is anything left after paying trustee and government fees, then the balance will be distributed among the ordinary creditors. Even among ordinary creditors, there are preferred categories for receiving payment, such as workers, municipalities, landlords, and government authorities.

The important conclusion is that you must act in a responsible, thoughtful, and discretionary manner to protect yourself. You must know with whom you are dealing. You must be aware of and be prepared to exercise your legal rights. You must not allow your accounts and your collections to fall behind.

19
Profiles and Pitfalls of Typical Projects

Over the years you get to bid on and do many different types of electrical installations. Chances are that you get to do certain types of jobs more frequently than others. You develop experience and expertise in certain types of work. Having this experience and expertise helps you to diminish the element of risk which is inherent in every electrical contract. At least you know what you may be up against and thus can take these factors into account. Your chances of succeeding on a project are greatly improved if you have staff and workers available with experience in that type of work. That experience will work for you. Otherwise, you will have to pay for them to learn from their mistakes.

Electrical work is electrical work, but not all projects can be handled in the same way. The different types of projects require particular expertise, particular tool requirements, and particular logistical planning. Therefore, when you decide to bid on or negotiate for a given project, make sure that you are in a position to handle it. Why end up buying a pack of trouble and a possible serious loss?

Some types of projects, such as hotels or speculative buildings, can confront you with continual design changes or with incomplete or poorly prepared drawings. These can result in lost worker-hours and poor productivity. Confusion in design, scarcity of details, interference between trades, slow answers to pressing questions, and unreasonable routing and fastening requirements are all destroyers of good productivity. Some projects are congested; others are spread out. Projects like sewage and filtration plants are complicated and of long duration. Other installations, such as general postal and mail-handling facilities, are subject to frequent changes and are interdependent with the work of other trades. In all of these projects there is the prospect of delays and wasted worker-hours.

Many contractors and construction managers have the reputation of being penny-wise. They provide too few hoists for crews and materials on highrise projects and thus condemn the trades to large losses of worker-hours. This is most evident around punch-in and punch-out time, break periods, and lunch periods and for material handling. Others work without a proper construction schedule, which results in a start-and-stop, uncoordinated work pattern. All of these factors should be taken into account when you are looking at a tender call or negotiating for a contract.

Between theory and practice, the construction industry often gets mired down in a swamp of problems. A good example is fast-track construction management. It sounds great in theory. A construction manager works with the design team, and the work is let out in packages before the total design is completed, thereby permitting a much more rapid start of the project. In practice, however, this can become a nightmare because of lack of proper coordination, lack of proper interfacing between the work of the various packages, and too many changes. The construction manager and the designer tend to lose sight of the limits that fixed prices impose on the scope of work performed by the various trades, and they may expect you to fill in gray areas of a contract without additional compensation.

Descriptions follow of some of the characteristics of various types of typical projects.

HIGHRISE OFFICE BUILDINGS

When you bid on the electrical work for a highrise building, you should be aware of the difference between a poured concrete structure and a steel-frame structure as far as your work is concerned.

In the case of a concrete-frame building, you become involved almost from the very beginning of the project in slab work and grounding. You will have to do more coordination during the early stages, in order to locate the required sleeves in the beams and accurately lay out the slab work. Early on, you will need released-for-construction drawings to permit you to do your coordination, because you are required to deliver some materials sooner than for a steel-frame building. For example, you will need the back boxes for the fire alarm and other systems much earlier if these have to be embedded in the cement. Shop drawings and measurements of the respective materials are required early enough that you can interface the installation of embedded conduit and boxes with the pouring schedule. Obviously, it is more economical to run as much as possible of your branch conduit, including the home runs, in the slabs. There are substantial economies in your labor costs if you can run conduits in the slabs, instead of on the underside of the slabs, since the latter operation results in costly fastening requirements. Once the slabs are poured and the jacks are removed, the floors are available to you.

In a steel-frame building, you must wait for the cement topping to be poured on the steel pans before the floors are available. Steel-frame buildings, however, allow more time for coordination and more flexibility for your installation. There is more scope to utilize prefabrication. If BX is specified for branch wiring, you can practically prefabricate this wiring ahead of time in your shop. You can then release the assemblies to the project as the floors become available. The result is a very productive way in which to work.

Serious problems in highrise construction invariably stem from inadequate hoisting facilities. Often there are too few hoists, no adjustable platforms to permit off-loading directly onto the hoists, and cabs that are too small to allow standard-size materials (such as 10-ft lengths of conduit on wagons) to be wheeled directly onto the hoists. Add to these deficiencies the lack of proper planning and scheduling for the use of the hoists, and you can end up with a waiting line of trucks and a pileup of subcontractors' materials. These inadequacies are even more serious when they pertain to the hoists for the crews. You can imagine the lines of workers waiting for the one hoist that very often services each project. Inadequate hoisting facilities can cost you 10 to 20 percent additional labor.

General contractors and construction managers very often insist on placing the lunchrooms and workers' shacks out of the way of the work. Time is lost during break periods to get to these shacks and then return to where the work is taking place.

During winter, the project may be inadequately heated in order to cut temporary heating costs. Labor productivity will suffer if that is the case.

Most highrise projects are on tight completion schedules, and you must ensure that your suppliers deliver their materials on time in order to cut down the movement and displacement of your workers. Your work schedule is disrupted when critical materials, which are necessary for an organized and continuous tempo of work, are not at the site on time. Other subtrades can upset your planned schedule of work if they have not coordinated their work with yours, or if they are late in starting an operation that affects your work. Very often the electrical installation is the third work done on the floor, after the sprinkler work and the ventilation work. Some trades, such as ventilation, drywall, and masonry, are known for cluttering up the floors with their materials. This can prevent the free passage of your scaffolding and workers. A general contractor who plans and coordinates the work and maintains a clean project will save you money.

There are two critical factors that you must control in order to cut down the lost time:

- Information must be available on time.
- Materials must be available on time.

After contractors have done a few highrise installations, the labor cost of doing this type of work will decrease, because they will know what to expect and how to program the work to reduce lost time. In every case, try to pick supervisors and crews who have the most recent experience in similar types of work. Learn lessons from previous projects, and take advantage of the experience gained from their methods and solutions. Try to ensure that your job supervisory personnel are compatible with and will get along with the personnel of the consultant and the general contractor. If the chemistry is right, the job will run smoothly.

Material handling will consume about 20 percent of your labor. You should assign a top person to look after this important function. You require a material-handling program to cut out unnecessary movement. This calls for adequate local containers, such as lock-up boxes, conduit wagons, and mobile material baskets. These will allow you to keep your central job storage down to a minimum. On highrise projects it is important to keep the installers working in place. Jumping around from operation to operation or from floor to floor will result in lost worker-hours. When you start an operation on a floor, keep at it until it is completed, before moving to another operation or another floor. This is a cardinal rule. Don't interrupt operations to handle materials. Use a separate crew for this purpose. Make absolutely sure that your suppliers cooperate with you to achieve your delivery, packaging, and identification requirements. This means delivering conduit and wire in accordance with job releases and interfacing truck deliveries with allotted hoisting time.

HOTELS

Hotel installations involve two main categories of work: the wiring of the rooms, which is a repetitive operation, and the wiring of the public areas, which is a custom-built, particular type of work. Your labor cost for wiring the rooms will depend on the types of walls which divide the rooms. In the majority of cases, the walls are gypsum board on metal studs. In some cases, however, the walls may be poured concrete and form part of the structure of the building. Your costs will be higher in the latter case.

When the walls are gypsum board on metal studs, the specified wiring is usually BX, which you can prefabricate in your shop. The installation of the drywalls is much faster than forming and pouring concrete walls. You can therefore do your roughing in faster and cut down the waiting time and lost time on this operation.

With concrete walls, your installers must be very careful to seal the boxes and conduits to prevent them from becoming blocked with cement. The water runoff from the poured concrete can penetrate into the conduits, where it will harden, particularly in elbows or low horizontal points. You will have to schedule your work to conform with the pouring schedule, and time may be lost waiting for the formwork to be completed.

Hotels are invariably affected by design changes, and you may be backed into a tight completion schedule to meet the opening date. You may be forced to accelerate and overstaff the job to achieve this. Usually, hotel tenders deal with the basic power and distribution for the building and the general wiring and lighting requirements. The public areas are subject to subsequent design and detailing. There are delays while these designs are formulated and approved. There are further delays until the prices for these change orders are negotiated and approved. The pressure will still be on, however, to complete the hotel on schedule, even though the changes will keep coming—up to and after the opening date.

Your work will be dependent on the progress of the other trades and on your ability to obtain timely information about items such as kitchen equipment, wall finishes, and decorative lighting. It is of utmost importance for the general contractor to produce and maintain an acceptable construction schedule. This is important as it relates to the pouring of the slabs and the installation of the gyproc walls, so that the rooms rapidly become available for roughing in and finishing. Repetitive operations are most effective when they proceed on schedule.

The work is usually much slower when it comes to public areas, such as the lobby, the ballroom, meeting rooms, bars, restaurants, kitchens, and recreational areas. These areas are treated in various decorative motifs, and it is quite likely that the required information will be slow in coming and subject to frequent changes.

Tendering on the electrical installation for a hotel can thus be very risky. If you have had previous experience in hotel contracts and have a staff who are familiar with this type of work, you will reduce the element of risk. An experienced staff will know what they are up against and will plan to utilize their resources in an efficient manner. They will be familiar with the many systems, such as dimming, fire alarm, sound, public address, and computers, which are common to this type of project. They will not be easily upset by the constant changes in design or the frequent demands of the general contractor, the consultant, and the interior decorator.

If the hoisting facilities are inadequate, you can expect traffic jams at the elevator, because of the large quantities of dry wall and other materials required for the rooms. You may experience difficulty in working in the rooms because such materials are piled up on the floors.

It is imperative that the design be set at a reasonable point in time. The drawings and design information should be available at an early date. The general contractor should maintain good coordination. On every project, the various parties solemnly promise to adhere to these principles; on most projects, such adherence is the exception, not the rule.

HOSPITALS

Hospitals are like hotels in that they consist of rooms and service areas, such as lobbies, restaurants, kitchens, laundry rooms, and operating rooms. They are very labor-intensive jobs as far as the electrical work is concerned. The chances are that you will tend to underestimate the labor cost for this type of work.

In each room, the quantity of electrical services will depend on the quantity of beds. Each bed usually requires the following services as a minimum:

1. Cord for the nurse's call
2. Power receptacle
3. Telephone outlet
4. Radio or TV outlet
5. Wall-bracket lighting fixture
6. Emergency circuit

Most rooms have false ceilings, and the wiring runs in the ceiling space in EMT conduit. The work in the rooms lends itself to prefabrication. All the room electrical services are collected in the corridor and run along there to the various system distribution points. For example, the nurse's-call wiring is run to the main nurses' call station on each floor. At this station there is also a doctors' paging system.

The electrical work in the operating rooms should be figured very carefully. The walls in these rooms are shielded, the floors are grounded, and the circuits are isolated. The isolating transformers for all circuits are usually mounted outside the room. Grounding and bonding of all the equipment is a critical requirement. The equipment is explosion-proof and has special fittings, and it is usually built of stainless steel. Your workers in these areas will have to be briefed and instructed regarding the very strict requirements for the electrical installation.

Hospitals require extensive emergency power for operating rooms and elevators and for emergency outlets and emergency lighting. You will have to do a substantial amount of coordination to prevent loss of worker-hours in areas such as kitchens, laundry rooms, operating rooms, laboratories, lobbies, and the morgue. Hospitals are usually very well designed and are not subject to many changes during construction.

You have to watch your cost in this type of project. The work is specialized. The location of much of your equipment must fit in with precise room or area layouts. The mounting heights vary and are critical in relation to the type and size of lab tables, beds, and operating-room layouts. Workers fastening materials onto an operating-room wall must take care not to puncture the shielding. The walls of the different areas are built and finished in various ways. Many of the switch and receptacle plates may have to be stainless steel. You will thus be confronted with exacting requirements for planning and coordination. Special attention must be paid so that material and equipment are ordered properly and delivered on time. Quality control of the work is a requirement for this class of work.

When you look back on the many problems that you have had on these types of jobs, it becomes painfully clear how simply they could have been prevented, if only you had thought about them ahead of time or had recognized them in time to take the necessary, logical steps to prevent them. But in most cases you didn't. The only way that experience works for you is if you learn from it and make the right moves in time. Hospitals are tough electrical jobs. You have to manage and control them very carefully in order not to suffer an overrun of your estimated costs, particularly of your labor costs.

INDUSTRIAL PLANTS

The plans and specifications for industrial projects are usually well laid out and complete. The industrial buildings themselves are well designed because they serve as the containers for the highly specialized production equipment and manufacturing operations that are housed therein.

When you obtain a contract for the electrical wiring of an industrial plant, your work may be affected by the lack of precise details regarding the production or process equipment. The process wiring very often comes out as a separate package, subsequent to the building contract. Often the owners will have prepurchased the major power equipment, the trays and Teck cables. These items were probably purchased from lists made up by the consultants. It is unlikely that job coordination requirements, optimum reel sizes, proper tagging, and required delivery dates have been taken into account. The jobsite storage of these prepurchased items, and the many problems which you will inherit when you come to install them, will involve you in more expense than would have been the case if you had purchased them.

Most industrial buildings are shells that sit on slabs poured on grade. You will not be able to do much work before the slab is poured. The roof drains have to be installed before the slab is poured. The production or process equipment is often scheduled for delivery when the slab has cured. You may thus be confronted with a floor that is cluttered just when the major portion of your work has to be done.

Industrial plants have high-bay areas. These present difficulties when it comes to installing your trays and fixtures, particularly if you have to do this work at the same time as the production equipment is being installed. The contractors who are installing the process equipment may require power for testing purposes sooner than you expected or scheduled for, and you may have to accelerate this phase of your work, with a resulting impact on your costs.

You may be required by the specifications to submit test reports and quality-control reports for the equipment which you supply. Make sure that these requirements are written into your purchase orders, and include arrangements for on-site tests by qualified personnel if such are required.

The roof must be completed and the building made watertight before you can move in your equipment. Otherwise, you may be forced to store and protect it. You may have to wait for external pipe structures and internal ventilation ducts and piping to be completed before you can install trays and Teck cables. You will require detailed information from the control and instrumentation contractors in order to be able to coordinate the installation of empty conduits or trays that you supply for the use of these subtrades. Without good coordination, you may be prevented from doing this work along with your other work in these areas.

The construction of the permanent roads to the project during the latter part of your contract may interfere with easy access to your work. The owner may occupy finished areas ahead of time, thus forcing your job personnel to go the long way round to get to the areas where they are working.

Some plants consist of many buildings and tank farms spread over a large area. You will require adequate vehicles to move your workers and materials around the site, as shown in Fig. 19.1. You should provide sufficient time clocks and arrange to position them so as to cut down punch-in and punch-out time. Industrial plants, such as those for pulp and paper and chemical processing, may be located in out of the way places, and you must look carefully into the available labor pool and working conditions in order to estimate the labor cost. You will require specific tools for specific types of work. Again, experience in this class of work is your key asset and your protection against the inherent risks.

Above all, make sure to coordinate your work with that of the other trades and with the process work. Try to install as much of your work as possible before the process equipment and conveyors are installed. Otherwise, your labor costs will increase. Color-code the conduit and wire, tag the equipment, and identify the circuits. Make sure that your purchaser schedules the materials to arrive on time. Keep control of the drawings, so that your installers are working with the latest information.

LARGE SEWAGE AND WATER-TREATMENT PLANTS

These projects consist of many buildings and structures connected by tunnels. The buildings house filter beds, storage basins, and pumping stations, and they cover a large area. A tremendous amount of concrete has to be poured. The construction schedule is, therefore, very

FIGURE 19.1   Material handling on a large project.

long—much longer than for most other types of projects. However, the greater part of the electrical work is usually done in the final 12 months.

These projects contain large quantities and large sizes of mechanical piping, and you will have to coordinate your work so as not to interfere with them. The installation of stainless steel piping and sophisticated mechanical equipment takes precedence over the electrical work. This means that you will have to run your trays in difficult areas, such as above the mechanical pipe racks. Yours will be the last trade to work in the tunnels, and you will have to deal with congestion of workers and materials due to the large quantities of mechanical piping. The ozonation and chemical treatment areas also consist of masses of mechanical piping, most of which will have been installed before you will be able to do your work there. You will have to work around and between this piping, and that will increase your labor costs.

The length of the tunnels, the large sizes of the filter areas, and the large amount of mechanical piping and equipment all make it very difficult to move materials around the site. There are many high-bay areas that are difficult to work in and require special scaffolding. You require expertise and experience to work in these high-bay areas and also to pull in the Teck cables on the long runs of tray. Your installers have to know how to set up the pulleys and snatch blocks for this critical operation.

You will not be able to start your work until the concrete finishing of the floors has been done. If this operation is delayed, you may be forced to accelerate to make up for the lost time. Condensation and dampness cause regular conduit to rust. You may therefore have to install large quantities of PVC conduits on these projects. Expertise and experience are required to bend and install the PVC conduit. Plastic conduit expands, contracts, and warps with changes of temperature. The fastening devices that you install have to allow this conduit to work and to move, so that the runs don't become distorted. Metallic conduit and equipment are specified to be treated to withstand corrosion. You will therefore have to coordinate and plan this work carefully, so as not to be delayed by the long delivery time that applies to this coated material.

Because the design is intricate, and is constantly being updated, you will require information and answers from the consultants which may be slow in coming. You may also have to wait for the mechanical shop drawings. Without the timely receipt of the above information and drawings, it will be difficult for you to plan and schedule your work. This may force you to go back to various areas or to pull in cables on trays where you had previously pulled in cables, thus requiring expensive duplicate rigging. Every time you go back to continue or complete an operation, you lose worker-hours.

On these spread-out types of projects you are required to work in many areas at the same time. You should therefore plan your work to cut down back-and-forth movement. Because of the large areas and long tunnels, you have to think out the logistics of placing job shacks and time clocks and of the material-handling program. You should store the materials as close to the work as possible. You will require equipment, such as forklift trucks, electric carts, mobile baskets, and conduit wagons, for efficient material handling. You will require material and tool storage boxes in sufficient quantities where the work is being done. Make sure to inspect them on a regular basis, to ensure that they don't become stocked with excess materials and tools. Good roads in and around the site are a necessity, to allow you to move materials to the entrance nearest to where the work is being done. You will require pickup trucks equipped to operate on propane so that they can be used inside the buildings.

Experience in these types of projects is vital for effective job planning and cost control. In these complex, spread-out jobs, worker-hours can be lost as easily as water through a sieve unless you exercise control. When you do undertake projects of this kind, take the time to study and determine where and how to run the trays to your best advantage, and how to program the work to prevent loss of time.

20
A Positive Professional Attitude

It is easy to get into a rut and to become locked into outdated attitudes and ways of doing things. You feel comfortable doing things in the accustomed way. Habits are hard to break. If you are locked into old, obsolete ways, you may get locked out of new business. Or you may get knocked out. You have to be flexible and open to new ideas and new ways. Harness the new ideas and management techniques to your good common sense and a caring attitude, and you will have the necessary thrust to advance in your field.

The need to increase productivity is a top priority in your industry. The programs to achieve this are often resisted by your workers because of an unspoken but nonetheless pervasive belief that the way to compensate for the cyclical nature of construction is to stretch out the work as a protection against layoffs. This attitude is particularly prevalent toward the end of a project, especially during slow times, when the prospects of being transferred to another job are slim. It is only natural for people not to want to work themselves out of a job. On many projects workers who want to produce may be under pressure, subtle or otherwise, to reduce the level of productivity in order to stretch out the work.

Don't fall back on convenient excuses or scapegoats to explain away these problems. A common excuse is to blame all the wasted worker-hours on the union steward or on the antiproductivity attitudes of some of the unions. Some of the blame surely lies there, but some of it lies with you and your associations. You have to become a better organizer and a better controller. You have to become a better communicator. Your industry has to address itself to the cyclical nature of construction and the resulting insecurities and attitudes. By building more security into the industry, you will be able to combat the naive and suicidal belief that the level of employment will somehow be increased by slowing down the work and reducing productivity.

If productivity is the thermometer, then your industry can be diagnosed as being sick. Your staff, your workers, and you depend on this industry for a good living. It is up to you, to all of you, to find the ways to achieve a lasting cure. To find a cure, a doctor must first diagnose the nature of the sickness. The responsibility of doctors is to cure their patients, not to kill them. This is the responsibility of the people involved in construction, as well. Most of your people participate in or watch some kind of sport. There would be no organized sports if players and management refused to respect or enforce the rules of the game. Without rules and respect for the rules, there can be no organized sports. The same is true of the construction industry.

The average contractor is a relatively small business person. You are very vulnerable to rising costs. You obtain most of your work by preparing fixed-price tenders. When you sign a contract, you are tied to your fixed price.

In order to prepare a fixed price now for work which will be done in the future, you have to be able to predict, with some certainty, what will be the cost of the materials and labor when the work is actually done. You cannot do this unless the owners, the consultants, the suppliers, and the unions will respect and live up to their agreements. The consultants have to be fair and flexible in the way they interpret the specifications. The suppliers have to respect and live up to their quotations. The unions have to respect the terms of their agreements and truly cooperate to improve productivity.

Some workers may think that they are beating the system by getting paid a maximum possible wage for a minimum effort. Those workers who consciously refuse to put in a good day's work are contributing to weakening the industry. Not only are those workers who go along with concepts or directives to slow down the work hurting the contractor, they are hurting themselves.

Productivity should be part of a continuous dialogue between management and labor. This dialogue should be based on facts and not on recriminations. Examine your own failings and don't cover up what may turn out to be your own lack of proper planning, coordination, and control of the work. You should put more emphasis on increasing the quantity and quality of the educational programs for your supervisors, foremen, and workers. These programs should concentrate on explaining the nature and ingredients of cost, labor units, planning, and material handling.

Many engineers and architects would be surprised, and perhaps indignant, if they were told that their designs and their attitudes often contribute to low productivity. Yet this is a fact. The design sector and the construction sector rarely collaborate during the design stage, and neither sector is sensitive to or cognizant of the problems of the other. The practical field experience of the construction sector is not reflected sufficiently in the design. Good labor-saving ideas proposed by contractors are often suspected as an attempt to cheapen the job.

The marriage of low bids on a competitive job often results in an adversary rather than a cooperative job climate, in which the engineer is expected to act as a police officer. The need to improve productivity should be an important concern in the design stage as well as in the construction stage. You are all in it together and must cooperate and coordinate your plans to solve design and installation problems, to speed up approvals, and to look at labor-saving proposals in a fair and flexible manner. In other words, management, labor, and the designers should be aware of the practical problems of the industry and consciously incorporate this awareness into their everyday activities.

You require a better public-relations program. This program should examine the way you and your industry are perceived by the public, the owners, and the consultants. The industry in general is doing an exemplary job, considering the quantity of risk and the magnitude and diversity of problems that you habitually take in your stride. You deserve a better word and a better image than you often get. You are a professional and you should insist on being treated as such, with respect, with consideration for your point of view, and with recognition of your rights.

Your intent in dealing with your customers should be positive and enthusiastic. You should make every effort to be as helpful as possible. Short of accepting unreasonable demands or uncalled-for or unnecessary costs, you should be very willing to help your customers solve their problems and achieve their aims. The reaction of some people when asked by a customer to do something that is out of the ordinary or that calls for extra thought and effort is to be negative or to refuse outright. Within reason, put yourself out. A positive, helpful, and enthusiastic reaction will win friends and improve the job climate.

On every jobsite you are part of a construction team. Your presence should be felt on this team as constructive and knowledgeable. There are, too often, components of this team who are indecisive, who have no plan, who jump around without a schedule. Don't contribute to the confusion if in fact it exists. Do your homework, deal with the facts and the real issues, and make sure that as much as possible your input is positive and to the point.

You should insist on a professional attitude to the work on the part of your personnel. There is far too much sloppiness, untidiness, and lack of concern in the industry. These invariably result in loss of worker-hours when you are required to clean up the deficiencies at the end of a job. You should train your personnel to check out and clean up the deficiencies on a continuous basis before the end of the job. The circuiting should be checked out and identified at the time the work is done, when the information and facilities are readily available. Many worker-hours are lost in correcting mixed-up circuiting, long after the work was done, particularly when there are staggered patterns in the lighting design, or sequence switching. Information that is readily available to the installers while they are doing their work, so that they can easily make the necessary corrections at the time, can become very complicated and time-consuming to obtain when a clean-up crew has to identify the wiring and repair the deficiencies months later.

At the end of most jobs you are left with deficiency lists pertaining to items that were not installed or repaired. Directories, nameplates, and other identification requirements have not been installed. Equipment has not been cleaned. Covers for panels or boxes have not been installed. Covers that are damaged, scratched, or missing have not been repaired or replaced. These deficiencies result from a sloppy attitude. All of these items should be looked after when the work is being done. Your workers should be aware that a sloppy, irresponsible attitude to deficiencies is intolerable.

The industry is paying tradespeople such as electricians in excess of $25,000 per year in wages, and you should deal with them accordingly, as professionals. You should expect a responsible and professional attitude from them in return. This requires communication. Good communication will improve motivation and will eliminate some of the time lost due to plain, ordinary ignorance of the information a worker needs to do a good job. The necessary data, information, and ideas, of the estimator, coordinator, purchaser, and project manager must reach the installers so that they are fully aware of events and feel part of the construction team.

The project managers and their foremen should plan the work in depth, so that the information, the materials, the tools, and a work plan are all available before the installers are committed to a work package. The foremen should spend the necessary time to review the work package with their installers before they start their work. The workers should have a picture of the operation and of the most economical fastening arrangements. When problems develop, they won't have to lose time waiting for the foreman to tell them how to handle them.

The actions required to save worker-hours are so evident, yet hours continue to be lost because of shortsighted, penny-wise attitudes with regard to job facilities, lunchroom placement, toilets, canteen timing, project lighting, cleanliness, the number of hoists, and the construction of loading ramps. If adequate facilities, proper planning, and good job management are supplied to the projects, they will pay back continuously in worker-hours saved. Unfortunately, many projects are shotgun marriages of the contractor and subcontractors who made the lowest bids. This results in an adversary job climate, in which the parties are only interested in protecting themselves. This is a fool's paradise, and in the long run the project and all the contractors will suffer. The cry is "We can't afford to do all these things." Not so! With the high cost of a worker-hour, you can't afford not to.

APPENDIX
Conversion Factors

U.S. Customary (or inch-pound) units of measure have been used in this book. To convert to metric—now called SI (Système International)—units, use the factors listed below.

CUSTOMARY UNIT	SI UNIT	CONVERSION FACTOR
Length:		
inch	millimeter	Multiply by 25.4
foot	meter	Multiply by 0.305
mile	kilometer	Multiply by 1.609
Area:		
square foot	square meter	Multiply by 0.093
Mass:		
pound	kilogram	Multiply by 0.454
ton (2000 lb)	kilogram	Multiply by 907.03
Temperature:		
degrees Fahrenheit	degrees Celsius	Use the formula $°C = \frac{5}{9}(°F - 32)$

GAUGE NUMBER EQUIVALENCES	
AWG NUMBER	MILLIMETERS
8	3.26
10	2.59
12	2.05
14	1.63

For more detailed information, refer to the following:

Metric Practice Guide, CSA Standard CAN 3-Z234.1, Canadian Standards Association, 178 Rexdale Blvd., Rexdale, Ontario M9W 1R3, Canada.

Metric Practice, ANSI/IEEE Standard 268-1982, American National Standards Institute, 1430 Broadway, New York, NY 10018.

Standard Metric Practice Guide, ASTM E380, American Society for Testing and Materials, 1916 Race St., Philadelphia, PA 19103.

Index

Acceleration, 69–71
  (*See also* Contract)
Accident report, vehicle, 163
Accounting, 83–84, 220–221
Additional work (*see* Change order;
    Work order)
Architect, role of, 96
Attitudes, positive professional, 275–
    279
Authority:
  abuse of, 7
  of architect, 56
  of engineer, 56
  exercise of, 5–7

Back orders, 148–149
  and short shipments, 148–149
Bar chart, 84–88
Billing, 253
  breakdown of contract for, 254–255
  financing expense, 259
  payment clauses affecting, 260
  percentages completed for progress,
    256
  system for progress, 256–259
Break periods, 16, 18
  time consumed by, 36
Buying (*see* Purchasing)

Cable and wire check list, 133, 143
Change order, 41–47
  cost of, 35–38
  estimating, 45–46
  impact of, 42
  quotation, 35–39, 203–206
    estimate sheet, 204
    record, 194, 197–198
    summary sheet, 205
  recovery of real cost of, 44–46
  types of: beneficial, 43
    corrective, 42
    impact, 43

Claim, 61–76
  documentation of: claim file, 63–64
    daily log book, 64
    foreman's weekly project report, 64–
      65
    job expense report, 67–68
    productivity check, 68–69, 88–90
    record of worker moves, 74–75
    stoppage of the work report, 66–67
  file for, 63–64
  types of: acceleration, 69–71
    extended duration, 62–63
Cleanup, 16
Clerk, job office, functions of, 187–189
Codes, 240–251
  job expense, 242, 248
  labor, 242, 244–245, 249
  location, 247, 249–251
  material, 241–248
    accessory and finishing, 245
    conduit, 243
    fixtures and lamps, 246
    outdoor and remaining, 248
    power and distribution equipment,
      246
    raceways and nonmetallic piping,
      244
    systems equipment, 247
    wire and cable, 243
Completion report, job percentage, 192–
    194
Computer:
  billing system, 253–259
  coding system, 240–251
  for cost control, 232, 237–240
  for estimating, 26–27
  input-operation report, 237
  on the jobsite, 235–236
  management and control functions,
    233–234
  output-status report, 238–240
  programming, 233–235
  role of, 231–232

Consequential damages, 260
Contingency, 32
Contract:
  acceleration of, 69–71
  breach of, 57
  changes to, 41–47
  cost-plus, 213–230
  damages: consequential, 260
    due to changed conditions, 49–55
    estimating, 57, 63–76
    job factors causing, 58–60
    notice regarding, 57–58
  definition of, 55
  delays and disruptions, 49–51
  extended duration, 62–63
  factors, 24
  notice provisions, 57–58
  rights and obligations, 61
Contract manager, 6, 214
Control (*see* Management and control
    functions)
Coordination, 206–208
Coordinator, 207–208, 214, 215
  (*See also* Contract manager)
Correspondence, 57–58, 194
Cost:
  breakdown of, 29–39
    direct job expense, 30–32, 39, 67–68
    general overhead, 33–34
    labor, 19–23, 36–38
    material, 21–23, 30
  of change orders, 35–38
  conditions influencing, 20, 23–25, 49–
    54
  control of, 3, 50, 79, 86–90, 91–103,
    106, 127
  estimated, 19–27
  of tools, 166–171
Cost-plus contract procedures:
  approvals, 215
  coding, 221
  forms, 222–226
  inventory and control of tools, 219
  material storage and handling, 217
  processing of invoices, 220
  purchasing, 215
  receiving of materials, 217
  return of materials, 218
  stores and warehousing, 217
  (*See also* Contract)
Crew (*see* Work force)

Daily labor report, 225
Damages (*see* Contract, damages)
Debit note, 230
Delay:
  claim, 49–76
  compensation for, 61–76
  factors causing, 53–54, 58–59
  impact on productivity, 49–54
Discipline:
  aims of, 3–7, 77–79
  as control, 7–8, 88–90
Disruption of work (*see* Delay)
Drawings:
  as-built, 209–210
  coordination, 206–208
  record of, 195–196
  shop, 208–209

Engineer, role of, 96
Equipment:
  record of changes to, 198
  rental report, 162
  (*See also* Material)
Estimate, role of, 27, 100–103, 237–238
Estimating:
  with computers, 26–27
  factors, 22–24
  job conditions, 20
  labor units, 19
  take-off items, 21–22, 29–35
Expense, direct job (*see* Cost, breakdown
    of)
Expense code, 242, 248
Experience, learning from, 9–10
Extended duration, 62–63
  (*See also* Claim; Contract)
Extra work (*see* Change order; Work or-
    der)

Fabrication, shop (*see* Prefabrication)
Field force report, 227
Files, job, 194–198
  change order, 194
  claim, 195
  general, 194
  labor, 195
  material, 194
  supplier, 194
Fixture check list, 136–137, 142
Forms (*see* Paper work forms)

Guarantee, 32

Hiring procedures, 79–82
Hospitals, 268–269
Hotels, 266–268

Identification:
  and marking, 210–211
  and tagging, 124–126
Industrial plants, 269–271
Instruction manuals, 210
Invoices, processing of, 220–221

Job expense, 30–32, 39, 242, 248
Job office clerk, 187–189
Job office procedures, 187–221
  filing, records, and reports, 189–198

Labor:
  burden of, 38
  cost of, 19–23, 36–38
  indirect, 11–15, 17–18
  management of, 5–9, 15–18
  material handling, 13–14, 17–18
  productive, 15–18
Labor code, 242, 244–245, 249
Labor multiplier, 36–38
Labor payroll:
  daily report, 214–215
  punchcard, 82
  weekly distribution sheet, 189–190
Labor report, 189–192
  field force, 227
  status, 239
  summary, 191–192
  weekly, 189, 191
Labor units, 11–18, 19–20
  factors influencing, 23–25, 88–90
Ladders (see Scaffolds, lifts, and ladders)
Layoffs, 83
Leadership, 6–7
Liens and privileges, 260–261
Lifts (see Scaffolds, lifts, and ladders)
Liquidated damages, 57
Log book, daily, 64

Management chart, 214
Management and control functions:
  account management, 232
  financial management, 232
  material management, 234
  project management, 233
Managing:
  computers in (see Computer)
  personnel manager, 77–78
  project (see Project manager)
Man-hours (see Worker-hours)
Manuals, instruction, 210
Marking, 210–211
Material:
  check list for, 133–137, 139–143
  cost of, 21–23, 30
  handling and storage of, 13, 152, 157–158
    secure, 154–155
  miscellaneous, 202
  ordering of, 12–13
    (See also Purchasing)
  receiving, 158–160, 217, 218
  receiving report, 159
  record of changes to, 198
  release form, 144
  release procedure, 137–145
  return/transfer of, 160, 218
  status report, 239–240
  storage of, 217–218
  storekeeper, 151, 153–154
Material codes, 241–248
Material requisition, 145–148
Manpower (see Work force)
Mechanics (see Work force)
Missing-tool report, 155
Morale, 97
Motivation, 6–7
Motor control center check list, 135–136

Notice provision clause in contract, 57–58

Office buildings, high-rise, 264–266
Operation (see Work operation)
Overhead, general, 33–34
  (See also Cost, breakdown of)
Overstaffing, 98–99
Overtime, 97–98

Panel check list, 141
Paper work forms:
  bar chart, 84–88
  for cost-plus contract, 222–226
  daily labor report, 225
  debit note, 230
  equipment rental record, 162
  job expense report, 67–68
  job requisition, 47
  labor multiplier, 36–38
  lost-tool report, 228
  missing-tool report, 155
  production sheet, 118–122
  productivity report, 89
  purchase order, 224
  purchase order memo, 223
  receiving report, 159
  release order, 144
  requisition, 147, 225
  return/transfer report, 161
  shop requisition, 123
  stoppage of the work report, 66–67
  subcontract, 149
  tool record, 229
  vehicle damage report, 162–163
  weekly project report, 64–65
  (See also Job office procedures)
Payment, progress, entitlement for, 253,
    254
  (See also Billing, system for progress)
Personnel:
  development of, 79
  evaluation of, 80–81
  factors, 24
  hiring of, 81–82
  policy for, 77–80
  (See also Work force)
Personnel manager, 77–78
Planning the work, 2
Plans:
  record of, 195–196
  revisions to, 41–43, 203–205
  study and check of, 11–12
Prefabrication:
  handling and tagging, 122–126
  and increased productivity, 99–100,
    105–107
  production sheets and requisitions for,
    116–122
  shop layout for, 107–113
  shop requisition for, 117, 123
  types of assemblies for, 113–116

Productivity:
  checking, 68–69
  effect of overtime on, 97–98
  effect of temperature on, 98–99
  factors influencing, 92–97
  impact of delay on, 49–54
  improving, 88, 91–103
    by control of slack time, 103
    by prefabrication, 99–100, 105–107
    by using optimum worker-hours, 91–
      92
  measurement of, 18
  reduction of, by overstaffing, 98–99
Productivity report, 89
Progress billing (see Billing, system for
    progress)
Progress payment (see Payment, pro-
    gress)
Progress reports, 237–240
Project manager, functions of, 1–10, 214,
    235–237
Project report, foreman's weekly, 64–66
Public relations, 277
Punch-in and punch-out, 4–5, 36, 82, 87
Purchase order, 149
  back order, 148–149
  check list for, 133–137, 139–143
  release, 137–145
  short shipment, 148–149
  for subcontract work, 149
Purchase order memo, 223
Purchasing:
  procedures for, 127–131, 133–137, 145,
    215–216
  review of materials prior to, 131–133

Receiving report, 159
Record:
  change-order quotation, 194, 197–198
  change order and work order, 198–
    208
  of changes to equipment, 198
  drawing, 195–196
  forms (see Paper work forms)
Release order, 144
Rental of equipment report, 160–162
Report:
  forms (see Paper work forms)
  labor, 189–192
  percentage complete, 192–194
  status (see Status report)

Requisition, job, 145–148
    form for, 147, 225
    form for materials from stores, 222, 225
Return/transfer report, 161
Risk, 1, 35
Runner, 153, 157

Scaffolds, lifts, and ladders, 167, 171, 175
Scheduling of work, 2–3, 84–88
Sewage treatment plants, 271–273
Shop requisition, 123
Shopwork (see Prefabrication)
Site factors, 23
Slack, control of (see Productivity)
Status report:
    budget, 238
    labor, 239
    labor per cost code, 239
    material, 239–240
    payment versus cost, 239
    planning and resource control, 240
    productivity per labor operation, 240
    quantity per material code, 240
    schedule, 239
Stoppage of the work report, 66–67
Storage facilities, 152, 217–218
Storekeeper, role of, 151, 153–154, 215–
    220
Subcontract form, 149
Switchgear and switchboard check list,
    133–135, 138, 140

Tagging and identification, 124–126
Team, two-worker, 97
Teamwork, 96–97
Temporary wiring:
    cost of, 185
    modular reusable components, 177–
        182
        light stand, 180, 182, 183
        power unit, 178–179, 181
        satellite plug-in unit, 179, 182
        typical requirements of, 182–185
        use of permanent installation for, 185–
            186
Termination of employment, 83
Timekeeping, 82–83
Tooling up, 14, 173–176
Tools:
    for conduit and raceway installation,
        168–169

Tools (Cont.):
    control of, 171–173, 219–220
    cost of, 166–171
    lost, report of, 228
    for material handling, 167
    missing, report of, 155
    problems with, 171–173
    record of, 229
    for wire and cable installation, 169
Transformer check list, 139, 142
Transmittal sheet, 196–197

Vehicle damage report, 162–163
Voucher, 200–203
    (See also Work order)

Warehouse, central, 155–157
    (See also Storage facilities)
Water treatment plants, 271–273
Weekly project report, 64–65
Work:
    activities in, 11–18, 93–94
    nonproductive, 11–15
    productive, 15
Work force:
    graph of, 84–88
    hiring of, 81–83
    personnel policy, 77–81
    sizing of, 7–8, 100–103
    timekeeping, 82–83
Work operation, 2, 9, 11
    broken down into activities, 11–18
    productivity levels, 93
Work order, 198–200
    voucher for execution of, 200–203
Workday:
    breakdown of, 93–95
    division of, 93–95
    typical, 93–95
Worker-hours:
    consumed by various activities, 93–95
    getting the most from, 88–90
    lost, due to relocation from one job to
        another, 86–87
    overrun of, 54–55
    reports on: summary labor, 191–192
        weekly labor, 189–191
        weekly labor distribution, 189–190
    targets, 100–103
Worker moves, 66–67, 74–76
Workers' shacks, 92, 94

About the Author

Sam Meland, P. Eng., has worked in the electrical construction field for the past 30 years. A graduate of McGill University in Montreal with a bachelor's degree in electrical engineering, he is general manager of Standard Electric Company Inc., a major Canadian electrical contracting firm, and a member of the National Electrical Contractors Association, the Engineering Institute of Canada, the Order of Engineers of Quebec, and the Canadian Society for Electrical Engineering.